# Innovation and Sustainability in Electric and Autonomous Mobility

Alberto Tomás Delso Vicente
*Department of Business Administration, Rey Juan Carlos University, Spain*

María García de Blanes
*Rey Juan Carlos University, Spain*

Óscar Aguado Tevar
*Nebrija University, Spain*

Luis Díaz-Marcos
*Nebrija University, Spain*

Margarita Núñez-Canal
*Nebrija University, Spain*

IGI Global
Scientific Publishing
Publishing Tomorrow's Research Today

| | |
|---|---|
| Vice President of Editorial | Melissa Wagner |
| Director of Acquisitions | Mikaela Felty |
| Director of Book Development | Jocelynn Hessler |
| Production Manager | Mike Brehm |
| Cover Design | Jose Rosado |

Published in the United States of America by
IGI Global Scientific Publishing
701 East Chocolate Avenue
Hershey, PA, 17033, USA
Tel: 717-533-8845 | Fax: 717-533-7115
Website: https://www.igi-global.com E-mail: cust@igi-global.com



Library of Congress Cataloging-in-Publication Data

LCCN: 2025021860 (CIP Data Pending)
ISBN: 9798337328027
eISBN:9798337328041
British Cataloguing in Publication Data
A Cataloguing in Publication record for this book is available from the British Library.

# Editorial Advisory Board

# Table of Contents

# Detailed Table of Contents



Although education is a key factor in achieving sustainable development, global inequalities in access and quality of education persist as major challenges. The education divide - the disparity in access to learning opportunities - is exacerbated by economic, technological and geographical factors, and particularly affects vulnerable communities. Artificial intelligence (AI) has emerged as a potential tool to reduce these inequalities, offering adaptive models that personalise instruction, optimise resources and promote educational equity. This chapter examines the role of AI in sustainable education, focusing on bridging the education divide through adaptive models. It explores innovative applications of AI in education, its advantages and challenges, and possible solutions so that integration into education systems is done ethically and avoids increasing discrimination.



Sustainability is a content of great importance to be implemented in the teaching-learning process from the early stages of the Spanish curriculum. This is indicated in its educational law, alluding to the concept of sustainability and its approach from the work of the Sustainable Development Goals (SDGs), in order to comply with the 2030 Agenda. Globally, almost 90% of people who attend primary education complete it, although the disparity in the regions where it is provided is still very uneven. This disparity has a number of causes, ranging from the availability of the necessary financial resources to be able to provide education worldwide for all girls and boys, to the lack of infrastructure in other cases. Thus, it is essential that a number of competencies that are essential for a successful life in society are acquired in primary education, and sustainability, through its Education for Sustainability (ESD), is one of them.



This chapter explores the role of artificial intelligence (AI) in transforming Physical Education (PE), highlighting its potential to personalize learning, enhance motor skill development, and improve motivation. AI tools such as motion analysis systems, virtual trainers, and gamified applications offer real-time feedback and individualized instruction, increasing student engagement and performance. Empirical evidence shows improvements in fitness, technique accuracy, and motivational climate when using AI-driven platforms. However, challenges remain, including teacher preparedness, ethical concerns about data use, and digital inequality. The chapter argues for a responsible and inclusive integration of AI in PE through training, infrastructure, and ethical guidelines, while emphasizing the need for further research into its long-term impact



Growing awareness of environmental issues has driven a change in the attitudes and behaviors of consumers, who are demanding more sustainable products and services. In this context, green or sustainable advertising emerges as a key tool for companies seeking to communicate their commitment to the environment and attract a public increasingly concerned about sustainability. The present research focuses on the descriptive analysis of green or sustainable advertising, exploring its characteristics, strategies, trends and challenges, with the aim of understanding its role in promoting more responsible consumption and building a more sustainable future. Some companies focus on highlighting the environmental benefits of their products, such as reducing emissions, using renewable energy or recycling materials. Other companies choose to communicate their sustainable business practices, such as implementing waste reduction programs, supporting conservation projects or committing to fair trade.

# Foreword

The contemporary world finds itself at a pivotal crossroads defined by two interlocking forces: the urgency of ecological sustainability and the unprecedented acceleration of technological innovation. Climate change, urbanization, and socio-economic inequalities are converging with artificial intelligence, edge computing, and autonomous systems, creating both opportunities and dilemmas that challenge our traditional models of governance, education, mobility, and communication. In this intricate scenario, the academic community is called upon not merely to observe, but to propose, design, and ethically evaluate the tools that will shape the future of our societies.

It is precisely within this critical juncture that the volume *Innovation and Sustainability in Electric and Autonomous Mobility* emerges as a highly significant and timely scholarly contribution. Edited by a group of researchers whose intellectual leadership spans across several disciplines, this book presents a comprehensive and pluralistic exploration of how innovation and sustainability intersect in shaping the technological, educational, communicational, and urban paradigms of the 21st century.

This is not a book that falls into the trap of techno-solutionism. On the contrary, the authors demonstrate a clear awareness of the ambivalences inherent in any technological transition. They understand that electric and autonomous mobility, smart city infrastructures, and artificial intelligence are not inherently sustainable or ethical; their societal value depends on the ways in which they are developed, implemented, and governed. As such, the volume offers a balanced, rigorous, and methodologically diverse examination of some of the most pressing issues of our time—grounded in empirical evidence, theoretical depth, and a critical lens that is, above all, constructive.

One of the most commendable aspects of this book is its intellectual architecture. The editors have succeeded in creating a narrative sequence that moves from the micro-level of educational practices and communication discourses to the meso- and macro-levels of urban mobility systems, smart infrastructures, and digital governance. This multiscalar perspective reflects a systems-thinking approach that is increasingly necessary to address complex global challenges. Readers will appreciate how the chapters are not isolated contributions but are conceptually and thematically interrelated, each offering a different angle from which to understand the innovation-sustainability nexus.

The chapters focused on education—particularly those examining the role of artificial intelligence in fostering equitable learning and the integration of sustainability in primary education—remind us that technological innovation begins not with machines, but with human development. The exploration of how physical education can benefit from AI tools to improve inclusion and motivation adds a fresh and underexplored dimension to the discussion. These contributions resonate with the growing recognition that the ecological and digital transitions must be accompanied by a profound educational transformation if they are to succeed in the long term.

Equally important are the chapters that scrutinize the communicative dimensions of sustainability. The distinction between genuine green advertising and strategic greenwashing is crucial in an age where corporate narratives can obscure more than they reveal. These analyses are not only relevant for marketing scholars but also for regulators, activists, and informed citizens who must navigate an increasingly saturated information environment. By examining the semiotics, strategies, and ethical implications of sustainability communication, the authors provide a toolkit for both critique and action.

The core of the book, however, lies in its treatment of electric and autonomous mobility and the broader framework of smart cities. These chapters delve into how new technologies can transform not just the way we move, but also how we live, perceive, and co-inhabit urban spaces. Particularly noteworthy is the chapter that applies the Viable System Model to urban mobility governance—an innovative attempt to reframe the city not merely as a space of transport but as a cybernetic system capable of adaptive self-management. Similarly, the discussions on smart infrastructure and edge computing go beyond the usual narratives of efficiency and delve into questions of energy justice, equity, and democratic oversight.

The inclusion of empirical studies—such as the analysis of payment preferences in carsharing services based on machine learning models—further anchors the book in real-world data. These studies are essential for exposing the hidden patterns of digital exclusion and economic inequality that often accompany innovation. By shedding light on these patterns, the authors challenge the assumption that technological progress is neutral or universally accessible.

Another strength of this book lies in its geographic and disciplinary diversity. From case studies in Spain to global urban benchmarks such as Singapore and New York, the chapters offer comparative perspectives that enrich the reader's understanding of context-sensitive solutions. The contributors bring expertise from fields as varied as education, communication, urban planning, engineering, sustainability science, and information technology. This interdisciplinarity is not decorative; it is foundational to the book's epistemological position. In a world of interconnected systems, siloed thinking is no longer sufficient.

From a policy standpoint, the volume aligns well with the strategic goals of the European Green Deal, the UN Agenda 2030, and Horizon Europe's missions related to climate-neutral cities and digital innovation. It provides actionable insights for decision-makers at multiple levels—from local administrators and educators to international institutions. Moreover, its academic rigor and methodological transparency make it a valuable resource for postgraduate courses, research programs, and professional training initiatives.

As a scholar who has spent the last two decades studying the socio-technical dynamics of urban innovation and sustainability, I consider this volume a model of what interdisciplinary research should look like: bold in its scope, nuanced in its analysis, and committed to the public good. It does not offer definitive answers—nor should it—but it provides the conceptual tools and empirical foundations necessary to ask better questions and to design more just and sustainable futures.

I enthusiastically recommend *Innovation and Sustainability in Electric and Autonomous Mobility* to researchers, practitioners, students, and policymakers alike. It is a book that informs, challenges, and inspires. In times of planetary uncertainty, it reminds us that innovation must be more than newness—it must be relevance, responsibility, and renewal.

**Marisol Carvajal Camperos**
*Faculty of Social Sciences, Rey Juan Carlos University, Spain*

# Preface

## I. OVERVIEW OF THE SUBJECT MATTER

The current volume, *Innovation and Sustainability in Electric and Autonomous Mobility*, emerges in response to a crucial and increasingly urgent academic and societal imperative: to understand, guide, and critically reflect upon the intertwined dynamics of technological innovation and sustainability across various domains of human activity. The book is conceived not as a monographic treatment of a single discipline, but as a mosaic of interdisciplinary contributions that collectively illuminate how emerging technologies—particularly artificial intelligence, edge computing, Internet of Things (IoT), and autonomous mobility—are reshaping the foundations of our societies in economic, social, environmental, and educational dimensions.

This work encompasses ten original chapters that offer a panoramic, yet analytically rigorous, exploration of innovation and sustainability in contemporary settings. It examines how digital and ecological transitions are co-evolving in different sectors, including formal education, urban mobility, corporate communication, tourism, smart infrastructure, and digital energy management. By bringing together voices from multiple disciplines and geographies, the book fosters a pluralistic understanding of the complex feedback loops between technological change and sustainable development.

The conceptual nucleus of the volume revolves around the belief that innovation must be both technologically efficient and socially responsible. In particular, the book problematizes the assumption that all technological advancement is inherently beneficial. Instead, the authors critically assess the unintended consequences of innovation—such as digital exclusion, ethical risks in AI adoption, socio-economic inequalities in mobility access, and the prevalence of greenwashing in corporate sustainability narratives. In doing so, the volume does not simply describe trends, but provides interpretative frameworks, empirical evidence, and actionable recommendations that can be adopted by policymakers, educators, urban planners, industry leaders, and academic researchers alike.

## II. WHERE THE TOPIC FITS IN THE WORLD TODAY

In the current geopolitical and environmental landscape, the pursuit of sustainability has evolved from a normative ideal into a strategic necessity. The acceleration of climate change, the rise of socio-economic inequalities, and the transformation of global production and consumption models have placed increasing pressure on all sectors to transition toward greener, fairer, and more resilient systems. Against this backdrop, innovation—particularly digital and technological—has become both a potential solution and a source of new challenges.

The integration of artificial intelligence in education, for example, holds immense promise for individualized learning, curriculum adaptation, and resource optimization. Yet, it also raises significant concerns about data privacy, ethical design, algorithmic biases, and the potential exacerbation of existing inequalities. Similarly, the adoption of autonomous vehicles and smart urban infrastructure introduces efficiencies in traffic management, emissions reduction, and energy use, but also invites debate on surveillance, equity in access, and the governance of urban data.

The chapters in this book capture this duality. They document not only the transformative power of innovation but also its ambiguous effects on societal well-being and ecological integrity. In a post-pandemic world where digitalization has become ubiquitous and where sustainability goals are integrated into corporate, governmental, and educational strategies, understanding these contradictions is more important than ever.

Moreover, the volume is situated at the intersection of several key agendas: the European Green Deal, the United Nations' Sustainable Development Goals (SDGs), the UNESCO Education for Sustainable Development framework, and Horizon Europe's missions on climate-neutral cities and digital transformation. In this sense, the book serves as a timely academic contribution aligned with global policy debates and real-world implementation efforts.

## III. DESCRIPTION OF THE TARGET AUDIENCE

This book is directed at a wide and diverse academic and professional audience, reflecting the breadth of the topics it addresses. On the academic side, the volume is especially relevant for scholars and researchers in the fields of:

- Sustainability studies
- Smart cities and urban planning
- Communication and media studies
- Educational technology and pedagogy
- Business and marketing ethics
- Electrical and systems engineering
- Artificial intelligence and digital transformation

Postgraduate students (Master's and PhD), postdoctoral researchers, and early-career academics will find in this compilation a robust source of theoretical insights, case studies, and methodologies to inform their own research. In addition, faculty members designing advanced seminars or interdisciplinary modules can use this volume as a reference text or complementary material.

From a practical perspective, the volume will also be of high value to policymakers, government agencies, public administration experts, urban mobility planners, educational policymakers, sustainability officers in corporations, and NGO leaders working at the intersection of technology and social innovation. By combining conceptual rigor with empirical data, and by offering a transdisciplinary perspective, the book contributes to decision-making processes that require both systemic thinking and critical evaluation of technological adoption.

In short, the target audience includes both creators and users of knowledge in the realms of innovation and sustainability. Whether they are theorists developing frameworks or practitioners applying them in real contexts, the readers of this book will gain insights into the implications, challenges, and potential solutions derived from the intersection of technology and sustainable development.

## IV. CHAPTER SUMMARIES AND THEIR IMPORTANCE

Each of the ten chapters included in this volume represents a distinct but interconnected contribution to the broader thematic frame of the book. Below is a detailed summary and justification for their inclusion:

### Chapter 1 – Education Divide and Technology: Can Artificial Intelligence Build More Equitable Learning?

This chapter explores the potential of AI technologies to bridge the global educational divide by personalizing learning experiences through adaptive systems, intelligent tutoring, and real-time feedback mechanisms. The authors also raise cautionary concerns about digital exclusion and propose ethical guidelines for inclusive implementation, making the chapter a vital contribution to the intersection of AI and social justice.

## Chapter 2 – Sustainability and Its Development in Educational Practices in Primary Education in Spain

Focusing on curriculum development and teacher training in Spanish primary schools, this chapter investigates how the principles of Education for Sustainable Development are being operationalized at the foundational level of the educational system. The chapter provides pedagogical innovations and empirical observations that underscore the transformative power of early education in fostering sustainability-oriented worldviews.

## Chapter 3 – The Integration of Artificial Intelligence in Physical Education: Innovation and Sustainability in Learning Performance

Blending technology with pedagogy, this chapter examines the role of AI in enhancing physical education through smart sensors, gamification, and real-time assessment. The discussion expands the application of AI beyond academic learning to include physical literacy and inclusive practices, emphasizing the importance of holistic educational innovation.

## Chapter 4 – Green or Sustainable Advertising: A Descriptive Analysis

This chapter traces the historical evolution and discursive strategies of green advertising. It examines how sustainability is communicated, commodified, and sometimes distorted in marketing practices. The authors offer a typology of green marketing strategies and assess their credibility, contributing to the literature on ethical communication and consumer behavior.

## Chapter 5 – Greenwashing as a Corporate Communication Strategy in Tourism Companies

Through case studies and empirical analysis, this chapter reveals how tourism companies manipulate sustainability narratives to construct favorable images without substantive environmental practices. The analysis underscores the importance of regulatory oversight and consumer awareness to counteract the deceptive practices of greenwashing.

## Chapter 6 – Innovation and Sustainability in Electric and Autonomous Mobility in Advertising and Media

This chapter investigates how electric and autonomous vehicles are framed in media discourses, focusing on the convergence of technological innovation and sustainability messaging. The findings highlight the impact of narrative framing on consumer perception and policy support, illustrating the role of media in legitimizing or questioning the transition to sustainable mobility.

## Chapter 7 – A Cybernetic Analysis of Urban Mobility through the Viable System Model

Introducing a systems theory perspective, this chapter applies the Viable System Model to the governance of urban mobility. It proposes new models for citizen participation, decentralized control, and dynamic feedback loops, offering a fresh lens for urban planners and policymakers seeking to enhance resilience and adaptability in transportation systems.

## Chapter 8 – Smart Cities and Infrastructure Industry

This chapter offers a comparative analysis of leading smart cities and their technological infrastructures. Drawing on real-world examples from Barcelona, Singapore, New York, and London, the authors explore both the opportunities and ethical dilemmas associated with data governance, surveillance, scalability, and equitable access in urban contexts.

## Chapter 9 – Edge Computing for Energy-Efficient Internet of Things

Focusing on energy management in IoT ecosystems, this chapter explores how edge computing can reduce energy consumption, enhance data security, and improve system performance. The discussion is highly relevant for engineers and digital sustainability advocates aiming to build low-carbon digital infrastructures.

## Chapter 10 – Prediction of Payment Method in Carsharing

Using machine learning and urban socio-economic data, this chapter analyzes the predictive relationship between neighborhood income levels and preferred payment methods in carsharing services. The results reveal digital inequalities in mobility access, offering new insights into the intersection of financial inclusion and shared mobility innovation.

## V. CONCLUSION: CONTRIBUTION TO THE FIELD

This volume offers a pioneering contribution to the academic and policy discourse on innovation and sustainability. By integrating multiple disciplines, geographic contexts, and methodological approaches, the book addresses both the promises and the pitfalls of contemporary technological advancement. Its central message is clear: innovation is not an end in itself, but a means to build more sustainable, equitable, and inclusive societies.

The combined analyses of educational transformation, green communication, smart urban systems, and digital infrastructure offer a comprehensive, evidence-based understanding of how different sectors can—and must—cooperate to align innovation with sustainability goals. Furthermore, the book elevates the importance of critical reflection, ethical foresight, and systemic design in shaping the technologies of the future.

In an era characterized by digital acceleration and ecological urgency, this book stands as both a scholarly resource and a call to action. It invites readers to rethink innovation not as a purely technical process, but as a cultural and ethical endeavor with far-reaching implications for the world we inhabit and the world we hope to create.

# Acknowledgment

The editors would like to extend their sincere gratitude to all the contributors of this volume, whose scholarly dedication, methodological rigor, and intellectual generosity have made *Innovation and Sustainability in Electric and Autonomous Mobility* possible. Each chapter reflects countless hours of research, critical reflection, and constructive dialogue, and we are deeply appreciative of the commitment demonstrated by all authors in meeting the highest academic standards.

We also wish to thank the peer reviewers for their careful evaluations and valuable recommendations, which have greatly strengthened the quality and coherence of the work. Their constructive feedback not only improved the individual chapters but also enhanced the thematic unity of the volume.

Special recognition is due to the institutions that supported this project—universities, research centers, and professional organizations—which provided an enabling environment for collaboration across disciplines and geographies. Their encouragement and facilitation of international academic exchange have been essential in bringing together diverse perspectives on innovation and sustainability.

Finally, we are grateful to IGI Global for their professional guidance throughout the editorial process, and to the readers who will engage with these pages. It is our hope that this volume serves not only as a repository of knowledge but also as a catalyst for further research, critical debate, and informed action toward more sustainable and equitable futures.

# Chapter 1
# Education Divide and Technology: Can Artificial Intelligence Build More Equitable Learning?

**Elena Fernández Gascueña**
https://orcid.org/0009-0001-6863-745X
*King Juan Carlos University, Spain*

**Enriqueta Villanueva-Montero**
https://orcid.org/0009-0005-2990-0883
*King Juan Carlos University, Spain*

**María García de Blanes Sebastián**
https://orcid.org/0000-0002-9169-3337
*King Juan Carlos University, Spain*

## ABSTRACT

*Although education is a key factor in achieving sustainable development, global inequalities in access and quality of education persist as major challenges. The education divide - the disparity in access to learning opportunities - is exacerbated by economic, technological and geographical factors, and particularly affects vulnerable communities. Artificial intelligence (AI) has emerged as a potential tool to reduce these inequalities, offering adaptive models that personalise instruction, optimise resources and promote educational equity. This chapter examines the role of AI in sustainable education, focusing on bridging the education divide through adaptive models. It explores innovative applications of AI in education, its advantages and challenges, and possible solutions so that integration into education systems is done ethically and avoids increasing discrimination.*

DOI: 10.4018/979-8-3373-2802-7.ch001

# INTRODUCTION

The implementation of artificial intelligence in the educational sector has the potential to significantly change the teaching and learning process. Adaptive AI models are built on algorithms that can quickly process large amounts of data and make decisions based on previous experience (Grassini, 2023). Technology has thus enabled the development of systems that allow for the personalization of material, rhythm, and level of difficulty according to each student's performance and learning style, improving inclusivity and knowledge retention (Slimi & Carballido, 2023). Luckin et al. (2016) have shown that AI systems can act as educational mediators by responding to the student's cognitive and emotional profile, resulting in more meaningful and engaging learning experiences.

One of the most significant applications of AI in the field of education is the development of intelligent tutoring systems, which simulate interactions with teachers. These systems can provide immediate feedback, identify misunderstandings, and suggest personalized learning exercises (Graesser et al., 2018). Likewise, AI allows teachers to obtain real-time analytical data on student development, resulting in more effective and evidence-based pedagogical intervention (Mohammed & Watson, 2019). These capabilities open up new options for improving quality, personalization, and equity in education. AI applications in education include virtual assistants and intelligent tutoring systems, as well as adaptive learning platforms that respond to student demands in real time. These technologies allow the content, pace, and difficulty level of activities to be adapted to suit each student's performance and learning style, thereby facilitating knowledge acquisition and enhancing quality (Zawacki-Richter et al., 2019). Apps can also automate administrative and assessment work, allowing teachers to focus on more pedagogically valuable activities (Cardona et al., 2023). Furthermore, in situations where connectivity is limited or there is a shortage of teaching professionals, AI can serve as a complementary solution to ensure access to quality content.

However, the implementation of these technologies must be followed by a rigorous examination of their ethical consequences, which include, among other things, data privacy issues, discriminatory biases, and the depersonalization of the educational process (Bozkurt, 2023; Ntoutsi et al., 2020). Furthermore, in this period of increasing global digitization, education faces a paradox: although technology provides new opportunities to democratize access to information, it also risks exacerbating disparities (Warschauer, 2004). The digital divide, defined as disparities in access, use, and appropriation of digital technology, is a structural impediment to the positive influence of digitalization on education systems. This disparity is visible not just in the availability of resources such as infrastructure and gadgets, but also in the quality of teacher training, students' digital literacy, and cultural adaptability to modern tools.

According to Moore, Vitale, and Stawinoga (2018), a lack of access to various devices and a poor internet connection has a negative influence on academic achievement, particularly in households with multiple students. In an analogous mindset, an OECD report (2021) underlines the significance of adopting future technologies, such as artificial intelligence, across all countries and social groups to avoid expanding education divide between contexts with different levels of digitalization. Both sources believe that the viability of an AI-powered educational change is predicated on the implementation of action frameworks that ensure systemic equity.

In this context, developing countries, as well as rural or marginalized regions, face difficulties in incorporating artificial intelligence into their educational institutions. The lack of connectivity, the shortage of technology resources, and the lack of specialist AI training for teachers all contribute to unequal implementation, potentially perpetuating or even deepening existing educational disparities. There is a large difference between high- and low-income countries, even within the latter category. Income levels differ by area, including Latin America and the Caribbean, Africa, and Southeast Asia. In such circumstances, the Lucas Paradox and the Innovation Paradox become apparent, underlining the reality that despite the potential benefits of investments in quality education—particularly in digital education—would bring to these regions, these investments do not materialize to the necessary magnitude. This is due to structural flaws that limit considerable progress toward eliminating poverty and inequality. It is therefore important to recognize that digitization is not a neutral process, but rather a socially conditioned phenomenon that requires active equality and accessibility policies to ensure its advantages reach all segments of the population.

The purpose of this paper is to address these challenges and answer the following research questions:

RQ1: How might adaptive AI models be used to narrow the education divide in different socioeconomic contexts?

RQ2: What are the main AI strategies to improve education through personalized learning?

RQ3: What difficulties and opportunities do AI offer in the realm of sustainable education?

RQ4. Can high- and low-income countries solve this issue in a balanced way?

This theoretical study aims to explore novel methods for adaptive models of artificial intelligence in education, as well as to examine the impact of AI in reducing the education divide. To achieve these objectives, we propose the following three stages. First, we will look at AI uses in teaching and learning. Second, we will look at how adaptive AI may help manage diversity. Finally, we will discuss the benefits and drawbacks of adaptive AI in terms of educational inequality. We will achieve this by evaluating the role of AI concerning SDGs.

## LITERATURE REVIEW

### Artificial Intelligence Applications

Technology has been used in education since the invention of computers, assisting both teachers and students with duties such as grades, research, and problem-solving (Slimi & Carballido, 2023). Technology-based teaching and learning systems have been developed to help students and teachers complete their objectives (Kabudi et al., 2021).

E-learning systems, such as content management systems (CMS), learning content management systems (LCMS), and learning management systems (LMS), emerged in the early twentieth century and have served as a supplement to traditional education (Ushakov, 2017). Kabudi et al. (2021) note that educational platforms like Canvas (Instructure, n.d.), Moodle (Moodle, n.d.), and Blackboard (Anthology Inc., n.d.) provide a diverse range of applications. These platforms allow for the incorporation of instructional materials that can be made available at any time, remote feedback, direct and effective communication, the use of different resources, content customization to fit unique needs, cost savings, and teamwork. This range of utilities fosters the development of collaborative skills, encourages student participation, and improves their overall experience.

Advances in digitalization have allowed for the development of systems capable of processing data at high speeds, corresponding to a huge qualitative leap in technological capabilities. These systems use data management algorithms that learn from the surroundings and prior activities, providing tailored reactions and decision-making abilities. The phrase "artificial intelligence" is used in the literature to refer to a group of digital technologies that can act independently and perform a wide range of jobs that increasingly approximate human intelligence (Slimi & Carballido, 2023). Generative artificial intelligence (GenAI) employs algorithms that can recognize patterns in data and produce new information based on similar rules (Pesovski et al., 2024; Slimi & Carballido, 2023). This information can be presented in various formats, including text, audio, photographs, and videos. Examples of this include ChatGPT and Perplexity. These programs may understand indicators offered within a context and respond appropriately, even merging multiple concepts to be considered creative. Furthermore, they are written and organized in narrative, schematic, or table format, and their versatility makes them quite close to human language (Grassini, 2023).

These characteristics, the ability to make decisions and to learn from previous data, can be used to design innovative proposals in the teaching-learning process, as they allow these systems to adapt quickly to the individual needs of students (Kabudi et al., 2021). Researchers believe that they will become increasingly important in education (Pesovski et al., 2024) because they make it easier to design student-centered learning strategies (Mohebbi, 2025). These strategies are underpinned by a constructivist approach to learning and the application of active methodologies. Through these methods, interaction, experimentation, and reflection, the student is placed at the center of the process (Córdova-Esparza et al., 2024). Active methodologies help students develop the ability to access, manage, and manipulate information. These skills are important because they help students build knowledge (Green & Donovan, 2018). Because they are interactive, they encourage the development of abilities such as cooperation and creativity (Córdova-Esparza et al., 2024). Furthermore, they have been shown to enhance student motivation and engagement (Ouyang & Zhang, 2024). Consequently, student-centered methodologies promote a more dynamic and independent teaching environment, empowering students to assume responsibility for their learning, and thereby transforming the roles of both teachers and students within the teaching-learning process (Martín-García et al., 2024).

AI can perform a wide range of tasks, which, when applied in active learning contexts, contribute to educational innovation (Slimi & Carballido, 2023). Specifically, the literature has described the use of artificial intelligence in the design of intelligent tutoring systems, which provide greater guidance to students in their work (Graesser et al., 2018); adaptive learning systems, which are platforms that can be adapted to the needs of students by adjusting the difficulty of tasks, presenting activities sequentially, and offering individualized feedback (Chiu et al., 2023); and recommendation systems based on data analysis that offer evaluative responses and suggestions for improvement (Mohammed & Watson, 2019). Consequently, technology plays an important role in establishing effective educational environments, enhancing the experience of both teachers and students, encouraging meaningful learning (Zawacki-Richter et al., 2019), and improving the quality of teaching.

Artificial intelligence (AI) can carry out activities that can be used in an educational context. We have classified these activities as follows: a) those related to the teaching-learning process; b) those related to evaluation; c) those related to administration; and d) those related to emotions.

a) According to the literature, teachers can use GenAI to create content for a variety of subjects and in a variety of formats, such as text, audio, and video, for the teaching-learning process. Furthermore, they are employed to provide instructional resources like tests and summaries (Bozkurt, 2023). Additionally, these resources give teachers important insights into the cognitive abilities, motivation, learning styles, and knowledge levels of their students. Teachers can use this information to choose appropriate tactics, offer individualized resources, and offer tailored support and direction (Huang & Chen, 2016). Furthermore, GenAI can be utilized to establish discussions and initiate projects, thereby fostering collaboration among students and enhancing their communication skills (Alam & Mohanty, 2022; Huang, Jing & Chen, 2016). AI has the potential to enhance the effectiveness of teaching by assisting with in-class management, task distribution, and verbal presentation of content (Alam & Mohanty, 2022; Gupta & Bhaskar, 2020; Huang, Jing & Chen, 2016). Additionally, it can facilitate the translation of educational materials (Kasneci et al., 2023). Teachers can incorporate feedback and comprehension questions or review questionnaires into teaching content (Jackson & Cossitt, 2015; Weston-Sementelli et al., 2018) to detect students' difficulties and diagnose errors, thus enabling them to offer feedback (Alam & Mohanty, 2022; Buchanan et al., 2021). Other GenAI uses, according to Salas-Pilco et al. (2022) and Saghiri et al. (2022), are virtual simulators, case studies, and laboratory experiments and selecting learning materials that are tailored to the individual needs of each student.

b) Assessment-related: Gen (JI) can assist teachers in assessing students' comprehension and critical thinking skills (Crompton & Burke, 2023) and academic integrity (Bond et al., 2024). It can also be used to grade essays, short answers, and standardized tests (Kasneci et al., 2023). The potential of AI to assess teaching capabilities is significant, as demonstrated by its capacity to evaluate the effectiveness of teachers' questioning techniques (Gunawan et al., 2021) and the quality of educational materials (Alkhalil et al., 2021), as well as ensuring students have access to and utilize these materials. This underscores its significant potential in the educational sector.

c) Administration-related: GenAI can be applied to decision-making related to student admission, seat reservation, and scheduling planning (Bond et al., 2024). It can also be used to predict students at risk of dropping out (Cardona et al., 2023) and to design retention strategies, such as recommending subjects or courses (Rabelo et al., 2024). Additionally, it offers the functionality to analyze scientific production and its impact, as well as to study collaboration between professors (Salas-Pilco et al., 2022). In addition, the system allows the generation of student prototypes and the evaluation of academic performance.

d) In the field of emotional intelligence, advanced AI systems are being integrated to detect emotions such as boredom or frustration in students, thereby assessing their engagement and regulating the content, as well as identifying those who may be experiencing difficulties (Gouseti et al., 2025).

In summary, the extensive applicability of AI is because it incorporates numerous options for designing personalized teaching. Acquiring competencies is easier for learners by allowing them to advance at their speed and receive tailored feedback and direction. Conversely, research indicates that when students receive feedback from an AI system, they feel better evaluated and are more conscious of their errors (Chiu et al., 2023). Research has also indicated that AI can promote student participation in the classroom; this is because some students may feel more comfortable interacting with technology than with teachers or classmates (Fu et al., 2020). Memarian & Doleck (2023) and Soong et al. (2020) underline that the individualized responses and personalized correction provided by AI teach students to be autonomous and to take responsibility for their own learning. In addition, AI-assisted active learning strategies can enhance creativity (Huang et al., 2021) and develop communication and problem-solving skills (Mokmin, 2020). Conversely, the adaptability enabled by AI has enhanced the teaching-learning process by making it more flexible, as it allows access to educational content at any time and place (Chan & Hu, 2023; Ntoutsi et al., 2020). In addition, it makes it easier for students with special educational needs to receive extra support that is tailored to their individual pace. This helps to reduce discrimination (Su & Yang, 2023). In fact, positive outcomes have been achieved through the creation of interactive learning settings related to students' emotional awareness, social skills, and cognitive development. Indeed, encouraging results have been achieved through the design of participatory learning environments related to cognitive development, social skills, and emotional awareness (Yang et al., 2024). In particular, data has shown that artificial intelligence is effective in helping to recognize emotions in autistic children (Kharbat et al., 2021).

On the other hand, the AI potential to perform routine tasks, including administrative and task evaluation, reduces teachers' workload, allowing them to focus on more valuable activities, thereby enhancing their efficiency (Chiu et al., 2023; Su & Yang, 2023). AI systems can assist in developing and enhancing teaching competencies. By providing objective feedback, these systems can make teachers feel less questioned and more inclined to adopt new approaches (Gunawan et al., 2021).

However, it is important to note that the applicability and advantages of AI applied to the teaching-learning process are not universally applied to all educational levels. In fact, despite the benefits promised by digital innovations, literature shows that the incorporation of technology in K-12 classrooms has had a limited positive effect on learning outcomes compared to traditional methodologies (Cheah et al., 2025). It highlighted how important teachers are in helping children at the lowest levels of the educational system to integrate technology. In this regard, Crawford et al. (2023) have shown that providing students with access to computers or the Internet does not affect their results. In practice, improvements in outcomes are more likely to be related to the use of educational software for personalized learning.

The use of digital technologies by K-12 teachers varies depending on whether they are teaching elementary or middle school students. In the latter case, it is used to personalize teaching through feedback, class and material organization, and assessment (Duan & Zhao, 2024; Kong & Yang, 2024). And it has also been used to promote autonomy and critical thinking (Memarian & Doleck, 2023). Nevertheless, the integration of AI into teaching and learning processes is much less prevalent in primary schools. As Moorhouse (2024) points out, it is mainly used for planning topics and preparing teaching materials. Díaz and Nussbaum (2024) found that platforms are used as tutoring systems, with students responding passively. Teachers at preschool and primary school tend to use digital technology more for supporting their classes or for delivering content or exercises, rather than for tasks such as assessment or feedback (Cheah et al., 2025).

In summary, GenAI has the potential to advance education at all levels. For example, students in elementary school can use this tool to improve their writing and reading abilities. This is achieved by correcting grammatical and syntactic errors and providing summaries of challenging topics. Furthermore, GenAI can generate questions that encourage critical thinking. At secondary and bachelor levels, AI can assist students in problem-solving and exercise development, as well as review questions, thereby helping them to contextualize and consolidate their learning. It provides step-by-step explanations to support their reasoning processes. At the university level, AI systems can assist in developing critical thinking, problem-solving, and research skills (Kasneci et al. (2023).

## Digital Divide: High-Income Countries vs. Low-Income Countries

Disparities in education and, more specifically, the digital divide, are more pronounced in low-income countries than in high-income countries. From this perspective, technology represents an additional form of marginalization, one more, of historically disadvantaged people and countries (Helsper, 2021). Limited access to new technologies in certain regions negatively impacts both teachers and students, thereby perpetuating socioeconomic differences. This problem is more prevalent in rural areas of low-resource countries, where access to quality internet connections is limited, and providing adequate digital skills training to teachers is impractical due to various structural limitations (Moreira et al., 2024; Villavicencio-Cedeño et al., 2024).

In line with the Sustainable Development Goals, and particularly SDG 4 on quality education, governments, institutions, and companies should prioritize education as a route to decent work (SDG 8) and social inclusion (SDGs 5 and 10), within the context of a suitable legal framework, particularly in the context of the emergence of artificial intelligence (Marwala & Mpedi, 2024; Yunga et al., 2023).

In high-income countries, the digital divide also affects older people, who are generally less able to adapt to new technologies than younger people. This phenomenon is known as the generational digital divide (Romero, 2020). It is difficult for them to find quality jobs that require knowledge of new technologies, even something as simple as accessing a job offer online can be difficult (Encabo, 2017). However, the largest divide is found in large cities in these countries, where population growth is driven by both international and internal migration (De Oteyza et al., 2021). From an economic perspective, people migrate from one place to another in the hope of improving their standard of living (Villaverde-Castro et al., 2017). In addition, young people are also attracted to the opportunities offered by "smart cities", which focus on the service sector (Honeker, 2018).

In the case of international migration specifically, people tend to move from low- and lower-middle-income countries to high-income countries. They arrive with widely varying levels of training. In higher education, however, the digital divide is less pronounced. The main issue they face is the lengthy bureaucratic process involved in having their qualifications recognized, which hinders their ability to find suitable employment in the host country. Conversely, individuals with a basic level of training exhibit a significant educational and digital divide (MPI, 2025).

Integrating this new population into the societies of the receiving countries is a key issue and one that usually presents challenges due to the new arrivals' differing socio-cultural backgrounds (Richmond, 1984). For migrants to successfully become familiar with their host country, the adaptation process should adhere to the standards set by the International Organization for Migration (IOM, 2025). Government measures supporting this should provide access to educational opportunities and reduce the digital divide within the migrant community (Cimar, 2022). It is important to recognize that migrants are seeking security and political stability, both of which are essential for ensuring a good quality of life and access to employment opportunities, and health services (Quijano, 2004). They aim to reduce social inequalities in their countries of origin, which can sometimes be exacerbated in their countries of destination during the absorption, integration, and adaptation process (Pinillos et al., 2019).

Conversely, the digital divide in low-income countries, particularly in the context of artificial intelligence (AI) and its correlation with the Sustainable Development Goals (SDGs), specifically SDG 4 (Quality Education), manifests itself in distinct ways depending on the geographical area. It is important to recognize the unique characteristics of Latin America, Africa, and Southeast Asian countries when making comparisons.

In the economic sphere, the Lucas paradox (Lucas, 1990) remains a valid theory: "The flow of capital from developed countries to developing countries is not as significant as would be expected to reduce poverty and inequality". This perspective is further reinforced by the finding that weak institutions and inefficient governance are the main causes of these nations' lack of economic growth, which impedes the achievement of SDGs 16 and 17 (Alfaro et al., 2008; Easterly & Levine, 2001).

The underperformance of investments in digital education, ICT infrastructure, and computer skills in developing nations is proof that Lucas' paradox still plays a role. Because of a lack of infrastructure, poor internet access, and insufficient basic computing power, these investments frequently do not yield the anticipated returns (Barua, 2020; Davies et al., 2019).

Nevertheless, only the most privileged classes gain from foreign investment in these nations, not the broader populace. Poor governance methods and high degrees of corruption are to blame for this (Zugaza et al., 2007). In this situation, not everyone has equal access to high-quality education, and the educational divide is growing until only the most privileged groups can afford it (Cano, 2014). According to the European Commission (2006), "the digital divide is the distance that separates people, business and territories in terms of opportunities to access and use ICTs". Therefore, in these countries where there is an education divide, there is also, by extension, a digital divide (Almenara, 2014).

In this perspective, the Innovation Paradox (Cirera & Maloney, 2017) is also pertinent. This phrase describes how, even though innovation is necessary for poor nations to flourish, they frequently do not embrace it or reap its benefits because they require effective public policies and initiatives. To achieve economic and social prosperity, they must build and preserve a strong capacity for innovation within a flexible governance framework, which they do not yet have. (Fagerberg & Srholec, 2008). The Innovation paradox has the immediate consequence of deepening the educational and digital divide. Technological advances tend to widen inequalities rather than reduce them unless they are accompanied by equitable conditions of access and appropriation.

Geographically speaking, the educational and digital divide is a complicated phenomenon that reflects and exacerbates gender and socioeconomic disparities, beginning with Latin America and the Caribbean. Limited access to information and communication technologies (ICTs) is one way that this divide is visible, and it has a detrimental effect on the standard of education and development prospects for large sectors of the population (Agüero et al., 2020). During the pandemic, schools' inability to deliver virtual instruction to students' homes was highlighted. Millions of students were still affected by school closures as 2021 came to an end. Some governments' efforts to expand digital learning programs have been limited in both scope and quality, which has had a negative impact on children with disabilities, children from low-income families, and—most importantly—those living in rural regions with migratory and ethnic groups. This circumstance highlights how urgently the digital divide must be closed to maintain educational continuity during emergencies (UNICEF, 2021).

The digital divide and access to technology in secondary school institutions were the subjects of a recent study conducted in Ecuador. The results showed significant differences in how technology resources are used in public and private institutions, especially by students from underprivileged backgrounds, and between urban and rural populations. Additionally, there is a difference in the digital proficiency of teachers between public and private schools, as well as between rural and urban locations. It is important to remember that, in both public and private schools, ten or more students share access to digital devices, and in certain situations, they are not even available. It was determined that to encourage investment in technology in public schools and teacher preparation in this field, governmental policies are required (Olvera et al., 2024).

The gender divide, and specifically the digital gender divide, should be emphasized in this context as a visible reality that represents women's unequal access, use, and appropriation of digital technologies (Vaca-Trigo & Valenzuela, 2022). Women are not as able to pursue technology-intensive careers as males because they do not have equal access to information and communication technologies (ICTs). The digital gender divide would be greatly reduced if this imbalance were addressed (Galperin & Arcidiacono, 2021).

According to the Commission for Latin America and the Caribbean (ECLAC, 2025), despite advancements in digital resources and connectivity, access to and utilization of digital tools remain inadequate and unequal, which limits the opportunities they present. One of the main causes of the structural disparities that will have an impact on the region's growth is the digital divide (Herrera et al., 2025).

Fundamental structural flaws that severely restrict effective access to information and communication technologies (ICTs), such as unstable electricity supply and inadequate telecommunications infrastructure, contribute to Africa's digital divide. It is important to remember, though, that this reality varies across the region. Although North African nations like Morocco, Algeria, Tunisia, and Egypt, as well as South Africa, have comparatively better connectivity, sub-Saharan Africa still has notably inadequate internet access (González & García, 2013).

The environment is characterized by high levels of corruption, institutional fragility, and geopolitical conflicts, all of which hinder the creation of equitable and sustainable educational programs and hence contribute to the persistence and generalization of illiteracy. Deep gender inequality, glaring sociocultural disparities, inadequate finance for teacher preparation, and inadequate educational facilities all contribute to the disparities in access to egalitarian education. The constraints about the absence of energy, internet connectivity, technological gadgets (hardware), and digital resources (software), along with the previously mentioned deficiency of training programs and ongoing assistance for educators, are especially pertinent (Ofosu-Asare, 2024).

Even while mobile connectivity has improved, stable and reasonably priced internet access remains a luxury for a large portion of the African continent, according to recent data from the International Telecommunication Union, a United Nations body (ITU, 2023). Furthermore, there is a lack of money for some initiatives aimed at increasing connectivity. The dependence on Official Development Assistance (ODA) as a fundamental element in the building of digital infrastructures is strengthened by this circumstance. In this regard, lowering structural inequality will be greatly influenced by the financial choices made by high-income nations. Every aspect of education is a priority that needs immediate attention and development (World Bank, 2021).

In Southeast Asia, the digital divide is a complicated problem that is impacted by both restrictive governmental frameworks and socioeconomic inequality. Lim (2017) investigates the effects of Indonesian governmental monitoring and censorship policies, emphasizing how these practices restrict multiple and egalitarian access to digital technology. The approach used by Gamage and Halpin (2007) provides an insightful viewpoint on the persistent difficulties with digital literacy and technology appropriation, especially in rural areas like Sri Lanka. Similar to this, the Asian Development Bank study from 2021 acknowledges the advancements in connectivity in the region while highlighting the continued existence of structural obstacles that prevent the development of a digital ecosystem that is really inclusive and equitable.

The rise of artificial intelligence (AI), a cutting-edge technology that has the potential to revolutionize social, educational, and productive systems, has further accelerated the digital transformation process on a worldwide scale in recent years. But the rise of this new technology has made already-existing differences between high- and low-income nations worse, especially in terms of the digital divide (Barro, 2025).

## DISCUSSION

It is clear that technology can help society progress and reduce inequality in all spheres, including education. In areas where there is a teacher shortage, intelligent tutoring systems that offer individualized assistance and support might make it easier for people to access high-quality content and suggest certain activities (Cardona et al., 2023). Similarly, the ability to create plans based on each student's knowledge and learning style might help lessen inequality in a variety of social contexts

(Kabudi et al.2021). In the same way, being able to use data to predict how well students will do or to check the risk of them leaving before finishing their studies means that we can do things to help them earlier, which makes it more likely that they will stay at school and continue their education in areas where there are not many opportunities (Chiu et al., 2023).

For example, the Zaya Learning Lab is an Indian platform that uses inexpensive hardware and AI-powered software to deliver tailored learning materials in places with poor internet connectivity. Additionally, tailoring the material to each student's unique needs helps teachers assess and analyze their development. AIR4Children is another Mexican project that uses open-source technologies to teach robotics and artificial intelligence to kids in low-income neighborhoods. The curriculum, which is based on Montessori education, aims to help Mexican youngsters develop STEM and computational thinking abilities at a young age.

However, a minimal degree of connection infrastructure and material resources, including both software and hardware, are necessary for the full potential of AI in education to be realized.

The context determines how AI affects the educational divide. It could solve the teacher shortage and make high-quality information more accessible in remote areas or nations with limited human and material resources. However, this would necessitate enough gadgets for teachers and students as well as at least passable connectivity. The issue is similar in terms of diversity because structural constraints may limit it (Holstein and Doroudi 2021). According to Warschauer (2003), these obstacles have the potential to worsen already-existing disparities and not only lessen but also maintain community divides.

On the other hand, the benefits of technology in education go beyond simply having access to sophisticated apps. On the one hand, it is important to acknowledge that for educators and learners to reap the benefits of artificial intelligence, they must be equipped with the skills necessary to use it efficiently. In the current digital era, literacy includes the capacity to use digital tools in addition to the more conventional abilities of reading and writing (Tiernan et al., 2023). According to Long and Magerko (2020), AI literacy is the collection of abilities that allow people to interact and work with AI agents, critically evaluate AI technology, and apply AI in a variety of contexts. To define the competencies that educators need to develop to incorporate AI into the classroom, UNESCO established an AI Competency Framework for Teachers (Cukurova & Miao, 2024). According to Cheah et al. (2025), one of the primary constraints faced by educators at all school levels is a lack of training. Chiu et al. (2023) have highlighted a significant problem: not all educators possess the necessary knowledge of AI technology. Many educators are unaware of the evaluation process and its rationale and are uncomfortable with how it operates and the recommendations it makes. Consequently, they are unable to provide students with adequate support.

Additionally, Cheah et al. (2025) draw attention to the dearth of training programs and the scant research output in this area, which is particularly problematic at the K–12 educational level. In particular, first- and second-order hurdles make it difficult for K–12 educators to incorporate AI into the classroom (Ertmer, 1999). While the latter is internal teacher variables like attitudes, beliefs, and confidence in using technology, the former is related to external restrictions like tools, resources, training, and policies (Wang, T. & Cheng, 2021). Research shows that instructors are more willing to embrace technology in the classroom if they think it will help them teach once resource scarcity has been resolved (Collie & Martin, 2024). For example, students and teachers may exhibit negative algorithmic bias, which shows up as mistrust and rejection of technology (Wang, S. et al., 2020). This mindset may have an impact on whether the system is used appropriately or for profit-making purposes.

It should be mentioned that studies on AI in education have looked at topics including individualized instruction, tools to enhance the teaching process, the benefits of utilizing AI, and how it affects student achievement. Nonetheless, research still hardly ever addresses the subject of ethics (Bozkurt, 2023).

The possibility of bias originating from the data itself is one of the dangers connected to the application of AI in education. This is because internet data is used to feed AI algorithms used in education, which can introduce biases that mirror societal norms (Ntoutsi et al., 2020). As a result, AI systems frequently struggle to appropriately reflect students' physical attributes and adjust to their cultural norms. According to Nye (2015), this is mainly because a large portion of the data that is supplied to these algorithms comes from Western, industrialized, highly educated, wealthy, and democratic nations. The design of global educational systems that ignore local sensitivities is a phenomenon known as AI colonialism in education (Holmes et al., 2018). As a result, learning materials that contain gender, racial, cultural, linguistic, ideological, or cognitive biases may be created when data is taken from the internet. When evaluating materials, educators need to be careful because biased data might be used to automate discriminatory content (Mouta et al., 2024).

Their usage to evaluate student profiles raises further ethical concerns because the conclusions and recommendations they make may impair the students' capacity for free will and decision-making (Burr et al., 2020). Furthermore, computers can aggregate data from many sources and infer information like sexual orientation, health, and political beliefs from fragmented data, thus data confidentiality is not guaranteed (Tundrea, 2020). Online learning environments also carry the danger of harassment, intimidation, or discrimination, as noted by Restrepo et al. (2019).

AI's ability to control emotions raises ethical questions about monitoring, the privacy of emotional experiences, the potential for emotion induction, and the connection between AI and conduct (Mouta et al., 2024). There are concerns about who owns and uses the cognitive and socioemotional data that is unintentionally provided by very young children (ages 4–10) interacting with robots (Mohammed & Watson, 2019).

The way ethical difficulties are handled in other disciplines is the source of other contingencies (Gorur et al., 2020; Karaca et al., 2021). Researchers agree that the number of programs offering ethical training to instructors and students is restricted and that the substance of these programs is also constrained (Ng et al., 2023; Schiff, 2022). To teach the ethics of AI with better transdisciplinary, the literature also acknowledges that ethical programs are mostly addressed from STEM and health disciplines and promote the inclusion of the humanities, arts, and social sciences in training plans (Dignum, 2021).

Concerns about the emergence of "cyborg identities" and their effects on educators and learners are also raised by the possibility of constructing avatars (Adams et al., 2022). According to Holmes et al. (2018), it has been proposed that most research on the effects of technology has been carried out either by the companies that developed the technology or by groups with financial interests and that these studies frequently use unrepresentative samples.

The potential for AI to take the role of teachers and lessen human involvement in instruction is another issue connected to its application in education. According to Mohebbi (2025), this problem could make students more reliant on technology, prevent them from developing their social and teamwork abilities, and even cause teachers to lose their authority (Miao et al., 2021).

Lastly, it should be mentioned that there are certain people who disagree with intelligent learning settings. The pedagogy associated with them has been connected to a new form of behaviorism called "Skinner's box" (Skinner, 2016; Watters, 2017), as acknowledged by Du Boulay (2019). This method's justification is that by using these methods, students take on a passive role, absorbing knowledge and following a set of instructions that are progressively more difficult and sequentially structured to help them reach complete competency. The author continues to address various criticisms, including the dehumanization of instruction led by computers rather than by instructors and the management of education by businesses interested in using students as products from which to gather data. In this context, other critiques have emphasized the decline in reading and writing abilities, the over-reliance on technological resources, and the decrease in educational activities (Pesovski et al., 2024).

In conclusion, there is still a long way to go before artificial intelligence systems in education reach their full potential and provide significant advantages to all stakeholders involved in the system.

Although we are not yet completely aware of the ramifications of employing a technology that has the potential to bring about significant changes, we are rapidly adopting it. One potential barrier is the digital divide, which can grow between people as well as between developed and underdeveloped nations, areas, and communities. AI integration has the potential to worsen already-existing social injustices. Leveraging the potential of AI while preventing its exacerbation of discrimination depends on several factors, including the conditions surrounding access to and use of technology, the availability of energy and connectivity, devices and platforms, and the digital literacy of teachers and students present in all contexts.

Regarding the first research question (RQ1), adaptive AI models offer a significant chance to decrease the educational divide in a variety of socioeconomic settings. Their effects, however, differ greatly across high- and low-income nations. Because there are strong IT infrastructures, qualified educators, and sophisticated digital integration regulations in place in the former, using AI is comparatively simple. On the other hand, as the literature review explains, low-income nations continue to face structural obstacles such as poor connectivity, subpar technology, inadequate teacher preparation, and institutional weakness (Ofosu-Asare, 2024; Moreira et al., 2024). Due to these constraints, AI does not always succeed in closing the divide and may instead exacerbate it in the absence of inclusive public policy.

AI must be viewed as a part of a holistic education equality plan rather than as a stand-alone remedy to solve these inequities. According to Holstein and Doroudi (2021), how AI affects inequality relies on the application setting as well as how human and technological resources are managed.

With regard to the second question (RQ2), the primary AI strategies for enhancing education through personalized learning encompass intelligent tutoring systems, adaptive learning, and predictive analytics platforms for student achievement. These tools are especially valuable for addressing student diversity, promoting autonomy, and improving academic engagement (Zawacki-Richter et al., 2019). However, personalization cannot be viewed in isolation. These tactics are only effective if teachers possess the necessary digital skills and if the necessary infrastructure is in place to support them. In many parts of Southeast Asia, Africa, and Latin America, this remains far from being accomplished (Olvera et al., 2024; ITU, 2023).

With regard to the third question (RQ3), AI represents both challenges and opportunities for sustainable education. On the one hand, it is a useful tool that can promote ongoing teacher training, maximize available resources, and lessen administrative duties, all of which increase the effectiveness of the educational system. On the other hand, there are serious ethical issues with algorithmic bias, privacy, student monitoring, and the possible dehumanization of instruction, especially if automated systems replace human intervention. These problems are made worse in situations where there is less institutional oversight or regulatory power, such as in certain underdeveloped nations with non-existent or inadequate data protection or ethical control laws.

Finally, the fourth question (RQ4) directly challenges the possibility of addressing these issues equitably between high- and low-income countries. According to research, there is not a single solution that works in this situation because of the wide differences in the opportunities, problems, and requirements. Debates in industrialized nations centre on issues like enhancing learning, the morality of data use, or the function of educators in intelligent environments. The priorities in developing nations, on the other hand, are more straightforward: guaranteeing connectivity, upholding the right to high-quality education, and educating educators in the fundamentals of digital literacy (Yunga et al., 2023; Marwala & Mpedi, 2024).

Thus, the Sustainable Development Goals—specifically, SDG 4 (quality education) and SDG 10 (reducing inequalities)—must be in line with the incorporation of AI into educational systems. According to the OECD (2021), for artificial intelligence to truly aid in the transformation of education, it needs to be supported by long-term public policies, investments in international solidarity, and an ethical viewpoint that prioritizes social justice and human well-being.

## CONCLUSION

One of the most significant areas of application for AI in education is the design of intelligent tutoring systems that simulate personalized interaction with a teacher. These systems can provide immediate feedback, identify misunderstandings, and recommend alternative learning paths. AI also allows teachers to access real-time analytical data on student progress, facilitating more effective and evidence-based pedagogical intervention.

This underscores the necessity for a high level of training, emphasizing a human-centered approach, the ethics of AI, and pedagogical expertise in utilizing AI for the development of teaching methodologies and practices (Ayyildiz & Yilmaz, 2023; Kong & Yang, 2024). From the students' perspective, the primary objective of AI interventions is to personalize learning, thereby enhancing participation and motivation, with minimal teacher presence (Adeyele & Ramnarain, 2024). The development of intelligent tutoring systems that simulate personal interactions with teachers is one of the most important uses of AI in education. Those systems can identify misconceptions, offer quick feedback, and suggest different learning strategies. AI enables more efficient and research-based pedagogical interventions by giving teachers access to real-time analytical data on students' development.

AI has the potential to transform education when integrated with pedagogical criteria and an approach to educational justice. When considering the application of artificial intelligence (AI) in the field of education, it is essential to understand that it cannot be analyzed in isolation, regardless of the socio-economic contexts that adopt it. Its real impact depends less on its technical sophistication and more on the conditions in which it is implemented: namely, students, teachers, and institutions operating within deeply unequal education systems. AI is not distributed equally across the world. Instead, it is incorporated in a fragmented way, influenced by factors such as access to connectivity, available infrastructure, and teacher training.

AI is not a neutral resource; it reflects and often amplifies existing inequalities. Therefore, it is not sufficient to merely reduce technological gaps in terms of access; it is imperative to critically examine the development of these tools, considering the objectives and values underpinning them. According to Holstein and Doroudi (2021), in high-income nations with integrated infrastructure, the danger is not a lack of infrastructure but rather its careless use, which feeds into the logic of automation and depersonalization of education. Transitioning from a superficial interest in technology to a deliberate integration that promotes critical thinking, student autonomy, and educational development is the challenge in this professional setting.

In low-income countries, however, structural constraints mean that AI cannot be fully exploited, even when there are international initiatives or accessible platforms. In these cases, the digital divide is not only technological but also cultural and pedagogical. It is challenging for many institutions to embrace more collaborative and interactive approaches because they are still based on outdated concepts. Inadequate devices unstable connectivity, and a lack of long-term public policies all contribute to these difficulties.

In light of these challenges, there is an urgent need to promote the development of contextualized technological solutions. The social, linguistic, and cultural dynamics of less developed regions are not often taken into consideration by the technologies produced in industrialized nations, which restricts their influence and reinforces what Tello (2023) refers to as digital colonialism. Promoting AI models developed from the requirements and expertise of the Global South's nations is essential to reversing this trend.

Therefore, policy choices that can address the systemic injustices that contribute to the digital divide will be essential for the future of AI in education. It separates people who may or may not critically adapt to technology in addition to those who have access to it. Many low-income nations have yet to give basic connectivity, rudimentary infrastructure, and teacher training priority, while high-income nations argue over the ethical or pedagogical applications of AI. These deficiencies limit information access and sustain other types of social exclusion, as the World Bank (2016) and ITU (2023) have noted.

Overcoming disparities that go beyond the availability of technology is a common challenge for regions like Southeast Asia, Africa, and Latin America. These problems are associated with inadequate investment in digital infrastructure, a lack of comprehensive governmental policies, and institutional inadequacy in fostering inclusive technical competencies. Accordingly, addressing the digital divide necessitates much more than just distributing gadgets; it is crucial to encourage a substantial appropriation of technology from the standpoints of social justice, sustainability, and educational parity. To achieve the desired results, a differentiated strategy is required that incorporates redistributive policies, strengthened human capacities, and participatory technological processes which respond to the actual conditions of each territory. It is imperative that we ensure AI is used as a tool for educational inclusion and does not become a new mechanism of exclusion.

## LIMITATIONS AND FUTURE LINES OF RESEARCH

Despite the implications and contributions, it makes, the work we present is not exempt from limitations that must be taken into consideration. Our thorough selection and review of the documentation using the WoS and Scopus databases, which are recognized as prestigious scientific sources, must acknowledge the researcher's subjective interpretation of data. Conversely, research on AI in education and its applications is very recent, so much remains to be learned about its effectiveness in an educational context.

This exploratory theoretical analysis is timely given the rapid advance of AI in education. Future research incorporating experimental methodologies to allow a comprehensive evaluation of the implementation of adaptive AI models and their effect on the educational process and the digital divide could enrich the literature. Research could combine quantitative analyses of the effectiveness of digital strategy application in teaching and learning at different levels and in different contexts. These analyses could focus on the experiences of students and teachers. It may also be advisable to develop specific indicators to measure the degree of technological equity and its correlation with the pedagogical use of AI tools.

Future research could examine the effectiveness of adaptive AI models and the impact of the digital divide on educational outcomes. Research in the future could also explore the ethics and fairness of integrating AI into education in more detail. We can look at issues like balance or priorities in teacher training because technological teaching systems need teachers who are experts in the subjects they teach and in didactics, as well as in using technology.

Research should focus on contrasting and obtaining sufficient empirical evidence about the real benefits of AI in education. To achieve significant improvements in student learning, it is essential to understand the most effective application strategies, the contexts and subjects in which they should be used, and the students for whom they are most beneficial. This work contributes to the debate on integrating AI into sustainable education by providing a theoretical basis for future research and policies that promote more equitable and accessible education.

Finally, this study proposes a comparative statistical analysis to provide empirical evidence of the differences in access to educational technology between countries with different levels of income. Adopting a quantitative approach, this study will seek to identify the key variables that need to be strengthened in each national context — such as connectivity, technological infrastructure, teacher training, and public investment — to make access to artificial intelligence in education more equitable. Unlike the theoretical approach adopted in this chapter, this empirical research will enable us to examine the factors that directly impact the digital divide. This will enable us to develop public policies that are more effective and tailored to the specific circumstances of each country.

## REFERENCES


Adams, C., Pente, P., Lemermeyer, G., Turville, J., & Rockwell, G. (2022). Artificial Intelligence and Teachers' New Ethical Obligations. *International Journal of Information Ethics, 31*(1). Advance online publication. DOI: 10.29173/irie483

Adeyele, V. O., & Ramnarain, U. (2024). Exploring the Integration of ChatGPT in Inquiry-Based Learning: Teacher Perspectives. *International Journal of Technology in Education, 7*(2), 200–217. DOI: 10.46328/ijte.638

Agüero, A., Bustelo, M., y Viollaz, M. (2020). Inequalities in the digital world?: Gender gaps in the use of ICT. https://doi.org/DOI: 10.18235/0002235

Alam, A., & Mohanty, A. (2022). Foundation for the Future of Higher Education or 'Misplaced Optimism'? Being Human in the Age of Artificial Intelligence. Paper presented at the *International Conference on Innovations in Intelligent Computing and Communications,* 17–29. https://doi.org/DOI: 10.1007/978-3-031-23233-6_2

Alfaro, L., Kalemli-Ozcan, S., & Volosovych, V. (2008). Why doesn't capital flow from rich to poor countries? An empirical investigation. *The Review of Economics and Statistics, 90*(2), 347–368. DOI: 10.1162/rest.90.2.347

Alkhalil, A., Abdallah, M. A. E., Alogali, A., & Aljaloud, A. (2021). Applying Big Data Analytics in Higher Education: A Systematic Mapping Study. [IJICTE]. *International Journal of Information and Communication Technology Education, 17*(3), 29–51. DOI: 10.4018/IJICTE.20210701.oa3

Almenara, J. C. (2014). Reflexiones sobre la brecha digital y la educación: Siguiendo el debate. *Inmanencia, 4*(2), 14–46.

Ayyildiz, P., & Yilmaz, A. (2023). A New Chapter Is Being Written about Writing Instruction: Instructional Leadership at K-12 Levels in the Age of Artificial Intelligence (AI). *Educational Policy Analysis and Strategic Research, 18*(4), 82–101. DOI: 10.29329/epasr.2023.631.4

Barro, S. (2025). The Digital Society Is More Unequal. Información Comercial Española, ICE. *Journal of Economics,* (938), 97–111. DOI: 10.32796/ice.2025.938.7890

Barua, S. (2020). Financing sustainable development goals: A review of challenges and mitigation strategies. *Business Strategy & Development, 3*(3), 277–293. DOI: 10.1002/bsd2.94

Bond, M., Khosravi, H., De Laat, M., Bergdahl, N., Negrea, V., Oxley, E., Pham, P., Chong, S. W., & Siemens, G. (2024). A Meta Systematic Review of Artificial Intelligence in Higher Education: A Call for Increased Ethics, Collaboration, and Rigour. *International Journal of Educational Technology in Higher Education, 21*(1), 4–41. DOI: 10.1186/s41239-023-00436-z

Bozkurt, A. (2023). Unleashing the Potential of Generative AI, Conversational Agents and Chatbots in Educational Praxis: A Systematic Review and Bibliometric Analysis of GenAI in Education. *Open Praxis, 15*(4), 261–270. DOI: 10.55982/openpraxis.15.4.609

Buchanan, C., Howitt, M. L., Wilson, R., Booth, R. G., Risling, T., & Bamford, M. (2021). Predicted Influences of Artificial Intelligence on Nursing Education: Scoping Review. *JMIR Nursing, 4*(1), e23933. DOI: 10.2196/23933

Burr, C., Taddeo, M., & Floridi, L. (2020). The Ethics of Digital Well-Being: A Thematic Review. *Science and Engineering Ethics, 26*(4), 2313–2343. DOI: 10.1007/s11948-020-00175-8

Cano, L. P. (2014). La corrupción y la ineficiencia en el gasto público local y su impacto en la pobreza en Colombia. http://hdl.handle.net/11445/712

Cardona, T., Cudney, E. A., Hoerl, R., & Snyder, J. (2023). Data Mining and Machine Learning Retention Models in Higher Education. *Journal of College Student Retention, 25*(1), 51–75. DOI: 10.1177/1521025120964920

Chan, C. K. Y., & Hu, W. (2023). Students' Voices on Generative AI: Perceptions, Benefits, and Challenges in Higher Education. *International Journal of Educational Technology in Higher Education, 20*(1), 43–18. DOI: 10.1186/s41239-023-00411-8

Cheah, Y. H., Lu, J., & Kim, J. (2025). Integrating Generative Artificial Intelligence in K-12 Education: Examining Teachers' Preparedness, Practices, and Barriers. *Computers and Education: Artificial Intelligence, 810*, 100363. Advance online publication. DOI: 10.1016/j.caeai.2025.100363

Chiu, T. K. F., Xia, Q., Zhou, X., Chai, C. S., & Cheng, M. (2023). Systematic Literature Rreview on Opportunities, Challenges, and Future Research Recommendations of Artificial Intelligence in Education. *Computers and Education. Artificial Intelligence, 4*, 100118. DOI: 10.1016/j.caeai.2022.100118

CIMAR. (2022). La brecha digital en el alumnado inmigrante: cifras según el informe PISA 2018. Congreso Internacional de Marketing, Comunicación y Estrategias. (CIMAR). https://2022.cimar.org.es/ponencia/la-brecha-digital-en-el-alumnado-inmigrante-cifras-segun-el-informe-pisa-2018/

Cirera, X., & Maloney, W. F. (2017). *La paradoja de la innovación: Las capacidades de los países en desarrollo y la promesa incumplida de la convergencia tecnológica*. Publicaciones del Banco Mundial.

Collie, R. J., & Martin, A. J. (2024). Teachers' Motivation and Engagement to Harness Generative AI for Teaching and Learning: The Role of Contextual, Occupational, and Background Factors. *Computers and Education. Artificial Intelligence, 6*, 100224. DOI: 10.1016/j.caeai.2024.100224

Córdova-Esparza, D., Romero-González, J., Córdova-Esparza, K., Terven, J., & López-Martínez, R. (2024). Active Learning Strategies in Computer Science Education: A Systematic Review. *Multimodal Technologies and Interaction, 8*(6), 50. DOI: 10.3390/mti8060050

Crawford, J., Vallis, C., Yang, J., Fitzgerald, R., & O'Dea, C. (2023). Artificial Intelligence is awesome, but good teaching should always come first. *Journal of University Teaching & Learning Practice, 20*(7), 1–12. DOI: 10.53761/1.20.01.01

Crompton, H., & Burke, D. (2023). Artificial Intelligence in Higher Education: The State of the Field. *International Journal of Educational Technology in Higher Education, 20*(1), 22. DOI: 10.1186/s41239-023-00392-8

Cukurova, M., & Miao, F. (2024). *AI Competency Framework for Teachers*. UNESCO Publishing.

Davies, I. E. E., Nwankwo, C. O., Olofinnade, O. M., & Michaels, T. A. (2019, November). Insight review on impact of infrastructural development in driving the SDGs in developing nations: A case study of Nigeria. In IOP Conference Series: Materials Science and Engineering (Vol. 640, No. 1, p. 012112). IOP Publishing. https://doi.org/DOI: 10.1088/1757-899X/640/1/012112

De Oteyza, M. O. G., Sosa, M. J., & Lara, M. C. F. (2021). Retos para la integración de las personas inmigrantes en las Smart Cities: El caso de Madrid. *Cuadernos Geográficos, 60*(1), 244–262. DOI: 10.30827/cuadgeo.v60i1.13733

Díaz, B., & Nussbaum, M. (2024). Artificial Intelligence for Teaching and Learning in Schools: The Need for Pedagogical Intelligence. *Computers & Education, 217*, 105071. DOI: 10.1016/j.compedu.2024.105071

Dignum, V. (2021). The Role and Challenges of Education for Responsible AI. *London Review of Education, 19*(1), 1–11. DOI: 10.14324/LRE.19.1.01

du Boulay, B. (2019). Escape from the Skinner Box: The Case for Contemporary Intelligent Learning Environments. *British Journal of Educational Technology, 50*(6), 2902–2919. DOI: 10.1111/bjet.12860

Duan, H., & Zhao, W. (2024). The Effects of Educational Artificial Intelligence-Powered Applications on Teachers' Perceived Autonomy, Professional Development for Online Teaching, and Digital Burnout. *International Review of Research in Open and Distance Learning, 25*(3), 57–76. https://doi.org/. v25i3.7659.DOI: 10.19173/irrodl

Easterly, W., & Levine, R. (2001). What have we learned from a decade of empirical research on growth? It's Not Factor Accumulation: Stylized Facts and Growth Models. *The World Bank Economic Review, 15*(2), 177–219. DOI: 10.1093/wber/15.2.177

Encabo, S. O. (2017). Digital Divide, Poverty, and Social Exclusion. *Labor Issues: Andalusian Journal of Work and Social Welfare*, (138), 285–313.

Ertmer, P. A. (1999). Addressing First- and Second-Order Barriers to Change: Strategies for Technology Integration. *Educational Technology Research and Development, 47*(4), 47–61. DOI: 10.1007/BF02299597

Fagerberg, J., & Srholec, M. (2008). National innovation systems, capabilities and economic development. *Research Policy, 37*(9), 1417–1435. DOI: 10.1016/j.respol.2008.06.003

Fu, S., Gu, H., & Yang, B. (2020). The Affordances of AI-Enabled Automatic Scoring Applications on Learners' Continuous Learning Intention: An Empirical Study in China. *British Journal of Educational Technology, 51*(5), 1674–1692. DOI: 10.1111/bjet.12995

Galperin, H., & Arcidiacono, M. (2021). Employment and the gender digital divide in Latin America: A decomposition analysis. *Telecommunications Policy, 45*(7), 102166. DOI: 10.1016/j.telpol.2021.102166

Gamage, P., & Halpin, E. F. (2007). E-Sri Lanka: Bridging the digital divide. *The Electronic Library, 25*(6), 693–710. DOI: 10.1108/02640470710837128

González Pascual, A., & García Jiménez, A. (2013). Internet y África: de la brecha a la esperanza digital. Redes, libertades y comunicación. http://hdl.handle.net/10115/15312

Gorur, R., Hoon, L., & Kowal, E. (2020). Computer Science Ethics Education in Australia - A Work in Progress. Paper presented at the 945–947. https://doi.org/ DOI: 10.1109/TALE48869.2020.9368375

Gouseti, A., James, F., Fallin, L., & Burden, K. (2025). The Ethics of Using AI in K-12 Education: A Systematic Literature Review. *Technology, Pedagogy and Education, 34*(2), 161–182. DOI: 10.1080/1475939X.2024.2428601

Graesser, A. C., Hu, X., Nye, B. D., VanLehn, K., Kumar, R., Heffernan, C., Heffernan, N., Woolf, B., Olney, A. M., Rus, V., Andrasik, F., Pavlik, P., Cai, Z., Wetzel, J., Morgan, B., Hampton, A. J., Lippert, A. M., Wang, L., Cheng, Q., & Baer, W. (2018). ElectronixTutor: An Intelligent Tutoring System with Multiple Learning Resources for Electronics. *International Journal of STEM Education*, *5*(1), 15–21. DOI: 10.1186/s40594-018-0110-y

Grassini, S. (2023). Shaping the Future of Education: Exploring the Potential and Consequences of AI and ChatGPT in Educational Settings. *Education Sciences*, *13*(7), 692. DOI: 10.3390/educsci13070692

Green, T. D., & Donovan, L. C. (2018). *Learning Anytime, anywhere through Technology Reconsidering Teaching and Learning for the iMaker Generation* (Hall, G. E., Quinn, L. F., & Gollnick, D. M., Eds.)., DOI: 10.1002/9781118955901.ch9

Gunawan, K. D. H., Liliasari, L., Kaniawati, I., & Setiawan, W. (2021). Implementation of Competency Enhancement Program for Science Teachers Assisted by Artificial Intelligence in Designing HOTS-Based Integrated Science Learning. *Jurnal Penelitian Dan Pembelajaran IPA*, *7*(1), 55–65. DOI: 10.30870/jppi.v7i1.8655

Gupta, K. P., & Bhaskar, P. (2020). Inhibiting and Motivating Factors Influencing Teachers' Adoption of AI-Based Teaching and Learning Solutions: Prioritization Using Analytic Hierarchy Process. *Journal of Information Technology Education*, *19*, 693–723. DOI: 10.28945/4640

Helsper, E. (2021). The digital disconnect: The social causes and consequences of digital inequalities.

Herrera, P., Huepe, M., & Trucco, D. (2025). Education and Development of Digital Skills in Latin America and the Caribbean. Economic Commission for Latin America and the Caribbean (ECLAC). https://repositorio.cepal.org/bitstreams/1bcc9786-a37c-4325-ba30-efe8b5f26022/download

Holmes, W., Bektik, D., Whitelock, D., & Woolf, B. (2018). Ethics in AIED: Who Cares? Paper presented at the *19th International Conference on Artificial Intelligence in Education,10948* 551–553.

Holstein, K., & Doroudi, S. (2021). Equity and artificial intelligence in education: will" aied" amplify or alleviate inequities in education? arXiv preprint arXiv:2104.12920. https://doi.org//arXiv.2104.12920.DOI: 10.48550

Honeker, A. (2018). Smart Cities and Youth Unemployment: Mobile Technology as a Tool for Connecting Labor Supply and Demand. https://repositorio.utdt.edu/handle/20.500.13098/12257

Huang, J., & Chen, Z. (2016). The Research and Design of Web-Based Intelligent Tutoring System. *International Journal of Multimedia and Ubiquitous Engineering*, *11*(6), 337–348. DOI: 10.14257/ijmue.2016.11.6.30

Huang, J., Shen, G., & Ren, X. (2021). Connotation Analysis and Paradigm shift of Teaching Design Under Artificial Intelligence Technology. *International Journal of Emerging Technologies in Learning (iJET), 16*(5), 73–86. https://doi.org/ . v16i05.20287DOI: 10.3991/ijet

International Organization for Migration. (2025). IOM - UN migration. Retrieved March 13, 2025, from https://www.iom.int/

Jackson, M., & Cossitt, B. (2015). *Is Intelligent Online Tutoring Software Useful in Refreshing Financial Accounting Knowledge? Advances in Accounting Education: Teaching and Curriculum Innovations*. Emerald Group Publishing Limited., DOI: 10.1108/S1085-462220150000016001

Kabudi, T., Pappas, I., & Olsen, D. H. (2021). AI-Enabled Adaptive Learning Systems: A Systematic Mapping of the Literature. *Computers and Education. Artificial Intelligence*, *2*, 100017. DOI: 10.1016/j.caeai.2021.100017

Karaca, O., Çalışkan, S. A., & Demir, K. (2021). Medical Artificial Intelligence Readiness Scale for Medical Students (MAIRS-MS) – Development, Validity and Reliability Study. *BMC Medical Education*, *21*(1), 10. DOI: 10.1186/s12909-021-02546-6

Kasneci, E., Sessler, K., Kuechemann, S., Bannert, M., Dementieva, D., Fischer, F., Gasser, U., Groh, G., Guennemann, S., Huellermeier, E., Krusche, S., Kutyniok, G., Michaeli, T., Nerdel, C., Pfeffer, J., Poquet, O., Sailer, M., Schmidt, A., Seidel, T., & Kasneci, G. (2023). ChatGPT for Good? On Opportunities and Challenges of Large Language Models for Education. *Learning and Individual Differences*, *103*, 102274. DOI: 10.1016/j.lindif.2023.102274

Kharbat, F. F., Alshawabkeh, A., & Woolsey, M. L. (2021). Identifying Gaps in Using Artificial Intelligence to Support Students with Intellectual Disabilities from Education and Health Perspectives. *Aslib Journal of Information Management*, *73*(1), 101–128. DOI: 10.1108/AJIM-02-2020-0054

Kong, S., & Yang, Y. (2024). A Human-Centred Learning and Teaching Framework Using Generative Artificial Intelligence for Self-Regulated Learning Development Through Domain Knowledge Learning in K–12 Settings. *IEEE Transactions on Learning Technologies*, *17*, 1562–1573. DOI: 10.1109/TLT.2024.3392830

Lim, M. (2017). Freedom to hate: Social media, algorithmic enclaves, and the rise of tribal nationalism in Indonesia. *Critical Asian Studies*, *49*(3), 411–427. DOI: 10.1080/14672715.2017.1341188

Long, D., & Magerko, B. (2020). What is AI Literacy? Competencies and Design Considerations. Paper presented at the *Proceedings of the 2020 CHI Conference on Human Factors in Computing Systems*, 1–16. https://doi.org/DOI: 10.1145/3313831.3376727

Lucas, R. E. (1990). VWhy Doesnpt Capital Flow from Rich to Poor Countries? V. V American Economic Review, 80(2), 92q96.

Luckin, R., & Holmes, W. (2016). Intelligence unleashed: An argument for AI in education.

Martín-García, J., Dies Álvarez, M. E., & Afonso, A. S. (2024). Understanding Science Teachers' Integration of Active Methodologies in Club Settings: An Exploratory Study. *Education Sciences*, *14*(1), 106. Advance online publication. DOI: 10.3390/educsci14010106

Marwala, T., & Mpedi, L. G. (2024). Social Protection. In Artificial Intelligence and the Law (pp. 119-141). Singapore: Springer Nature Singapore. https://doi.org/DOI: 10.1007/978-981-97-2827-5_7

Memarian, B., & Doleck, T. (2023). ChatGPT in Education: Methods, Potentials, and Limitations. *Computers in Human Behavior: Artificial Humans*, *1*(2), 100022. DOI: 10.1016/j.chbah.2023.100022

Miao, F., Holmes, W., Huang, R., & Zhang, H. (2021). *AI and Education: A Guidance for Policymakers*. Unesco Publishing.

Migration Policy Institute (MPI). (2025). Migration Policy Institute. https://www.migrationpolicy.org/

Mohammed, P. S., & Watson, E. '. (2019). Towards Inclusive Education in the Age of Artificial Intelligence: Perspectives, Challenges, and Opportunities. *Artificial Intelligence and Inclusive Education: Speculative Futures and Emerging Practices*, 17–37. https://doi.org/DOI: 10.1007/978-981-13-8161-4_2

Mohebbi, A. (2025). Enabling Learner Independence and Self-Regulation in Language Education Using AI Tools: A Systematic Review. *Cogent Education*, *12*(1), 2433814. DOI: 10.1080/2331186X.2024.2433814

Mokmin, N. A. M. (2020). The Effectiveness of a Personalized Virtual Fitness Trainer in Teaching Physical Education by Applying the Artificial Intelligent Algorithm. *International Journal of Human Movement and Sports Sciences*, *8*(5), 258–264. DOI: 10.13189/saj.2020.080514

Moore, R., Vitale, D., & Stawinoga, N. (2018). The digital divide and educational equity. *Insights in Education and Work, 1*–10.

Moorhouse, B. L. (2024). Beginning and First-Year Language Teachers' Readiness for the Generative AI Age. *Computers and Education: Artificial Intelligence, 610*, 100201. Advance online publication. DOI: 10.1016/j.caeai.2024.100201

Moreira, J. L. C., Aguilera, A. G. F., Solís, I. A. M., Dután, J. F. L., Centeno, M. V. C., & Taco, N. B. I. (2024). Reducción de la brecha digital en zonas rurales: soluciones tecnológicas para una educación

Mouta, A., Torrecilla-Sanchez, E., & Pinto-Llorente, A. (2024). Design of a Future Scenarios Toolkit for an Ethical Implementation of Artificial Intelligence in Education. *Education and Information Technologies*, *29*(9), 10473–10498. DOI: 10.1007/s10639-023-12229-y

Ng, D. T. K., Lee, M., Tan, R. J. Y., Hu, X., Downie, J. S., & Chu, S. K. W. (2023). A Review of AI Teaching and Learning from 2000 to 2020. *Education and Information Technologies*, *28*(7), 8445–8501. DOI: 10.1007/s10639-022-11491-w

Ntoutsi, E., Fafalios, P., Gadiraju, U., Iosifidis, V., Nejdl, W., Vidal, M., Ruggieri, S., Turini, F., Papadopoulos, S., Krasanakis, E., Kompatsiaris, I., Kinder-Kurlanda, K., Wagner, C., Karimi, F., Fernandez, M., Alani, H., Berendt, B., Kruegel, T., Heinze, C., & Staab, S. (2020). Bias in Data-Driven Artificial Intelligence Systems-An Introductory Survey. *Wiley Interdisciplinary Reviews. Data Mining and Knowledge Discovery*, *10*(3), e1356. DOI: 10.1002/widm.1356

Nye, B. D. (2015). Intelligent Tutoring Systems by and for the Developing World: A Review of Trends and Approaches for Educational Technology in a Global Context. *International Journal of Artificial Intelligence in Education*, *25*(2), 177–203. DOI: 10.1007/s40593-014-0028-6

OECD. (2021). Digital Education Outlook 2021: Pushing the Frontiers with AI. *OECD Publishing*. https://doi.org//arXiv.2104.12920DOI: 10.48550

Ofosu-Asare, Y. (2024). Developing classroom ICT teaching techniques, principles and practice for teachers in rural Ghana without access to computers or internet: A framework based on literature review. *The International Journal of Information and Learning Technology*, *41*(3), 262–279. DOI: 10.1108/IJILT-04-2023-0045

Olvera, E. Y. M., Bastidas, E. G. J., & Espinoza, G. J. M. (2024). Analysis of the Digital Divide and Access to Technological Resources in Secondary Education Institutions in Ecuador. *Ciencia Latina Multidisciplinary Scientific Journal*, *8*(2), 6698–6719. DOI: 10.37811/cl_rcm.v8i2.11086

Organisation for Economic Co-operation and Development (OECD). (2021). *Digital education outlook 2021: Pushing the frontiers with AI*. OECD Publishing., DOI: 10.1787/589b283f-

Ouyang, F., & Zhang, L. (2024). AI-Driven Learning Analytics Applications and Tools in Computer-Supported Collaborative Learning: A Systematic Review. *Educational Research Review*, *44*, 100616. DOI: 10.1016/j.edurev.2024.100616

Pesovski, I., Santos, R., Henriques, R., & Trajkovik, V. (2024). Generative AI for Customizable Learning Experiences. *Sustainability (Basel)*, *16*(7), 3034. DOI: 10.3390/su16073034

Pinillos, G., Masferrer, C., & Oropeza, N. (2019). Inequalities, Migration, and Borders: A Proposal for a Multidimensional Approach. *Notes on Migration and Inequalities*, *3*, 2–13.

Quijano, A. (2004). The Labyrinth of Latin America: Are There Other Ways Out? *Journal of Iberian and Latin American Research*, *10*(2), 173–196. DOI: 10.1080/13260219.2004.10426802

Rabelo, A., Rodrigues, M. W., Nobre, C., Isotani, S., & Zárate, L. (2024). Educational Data Mining and Learning Analytics: A Review of Educational Management in e-learning. *Information Discovery and Delivery*, *52*(2), 149–163. DOI: 10.1108/IDD-10-2022-0099

Restrepo, E. G. Y., Baldassarre, M., & Boticario, J. G. (2019). Accessibility, Biases and Ethics in Chatbots and Intelligent Agents for Education. Paper presented at the *EDULEARN19 Proceedings,* 8824–8833. https://doi.org/DOI: 10.21125/edulearn.2019.2196

Richmond, A. H. (1984). Adaptation and Sociocultural Conflicts in Immigrant-Receiving Countries. https://unesdoc.unesco.org/ark:/48223/pf0000062549_spa

Romero, A. M. M. (2020). The Generational Digital Divide. *Labor Issues: Andalusian Journal of Work and Social Welfare*, (151), 77–93.

Saghiri, M. A., Vakhnovetsky, J., & Nadershahi, N. (2022). Scoping Review of Artificial Intelligence and Immersive Digital Tools in Dental Education. *Journal of Dental Education*, *86*(6), 736–750. DOI: 10.1002/jdd.12856

Salas-Pilco, S. Z., Xiao, K., & Hu, X. (2022). Artificial Intelligence and Learning Analytics in Teacher Education: A Systematic Review. *Education Sciences, 12*(8), 569. DOI: 10.3390/educsci12080569

Schiff, D. (2022). Education for AI, not AI for Education: The Role of Education and Ethics in National AI Policy Strategies. *International Journal of Artificial Intelligence in Education*, *32*(3), 527–563. DOI: 10.1007/s40593-021-00270-2

Skinner, B. F. (2016). *The Technology of Teaching*. BF Skinner Foundation.

Slimi, Z., & Carballido, B. V. (2023). Systematic Review: AI's Impact on Higher Education - Learning, Teaching, and Career Opportunities. *Tem Journal-Technology Education Management Informatics, 12*(3), 1627–1637. https://doi.org/DOI: 10.18421/TEM123-44

Soong, R., Pautler, B. G., Moser, A., Jenne, A., Lysak, D. H., Adamo, A., & Simpson, A. J. (2020). CASE (Computer-Assisted Structure Elucidation) Study for an Undergraduate Organic Chemistry Class. *Journal of Chemical Education*, *97*(3), 855–860. DOI: 10.1021/acs.jchemed.9b00498

Su, J., & Yang, W. (2023). Unlocking the Power of ChatGPT: A Framework for Applying Generative AI in Education. *ECNU Review of Education*, *6*(3), 355–366. DOI: 10.1177/20965311231168423

Tello, A. (2023). On Digital Colonialism: Data, Algorithms, and the Technological Coloniality of Power in the Global South in *Inmediaciones de la Comunicación*, Vol. 18, no. 2.https://doi.org/DOI: 10.18861/ic.2023.18.2.3523

Tiernan, P., Costello, E., Donlon, E., Parysz, M., & Scriney, M. (2023). Information and Media Literacy in the Age of AI: Options for the Future. *Education Sciences*, *13*(9), 906. DOI: 10.3390/educsci13090906

Tundrea, E. (2020). Artificial Intelligence in Higher Education: Challenges and Opportunities. *14th International Technology, Education and Development Conference (Inted2020),* 2041–2049. https://doi.org/DOI: 10.21125/inted.2020.0644

UNICEF. (2021). *Reimagining Education in Latin America and the Caribbean.* https://www.unicef.org/lac/en/reimagine-education-latin-america-and-caribbean

Ushakov, A. (2017). Learning Content Management Systems in Flt: Canadian Experience. *International Journal of English Language and Literature Studies, 6*(1), 25–32. RePEc: asi: ijells:v:6: y:2017: i:1:p:25-32:id:531

Vaca-Trigo, I., & Valenzuela, M. E. (2022). Digitalization of Women in Latin America and the Caribbean: Urgent Action for a Transformative and Equal Recovery (No. 47940). United Nations Economic Commission for Latin America and the Caribbean (ECLAC). RePEc: ecr: col022:47940

Villaverde Castro, J., & Maza Fernández, A. J. (2017). Internal Migration in Spain, 2000–2014: Evidence from a Spatial Econometrics Analysis.

Villavicencio-Cedeño, M. M., Bowen-Anchundia, M. M., Jurado-Martínez, M. G., & Roger-Martínez, I. (2024). The Digital Divide in Student Education: Determining Factors, Educational Consequences, and Proposals for Its Mitigation. *MQRInvestigar*, *8*(4), 6641–6673. DOI: 10.56048/MQR20225.8.4.2024.6641-6673

Wang, S., Yu, H., Hu, X., & Li, J. (2020). Participant or Spectator? Comprehending the Willingness of Faculty to Use Intelligent Tutoring Systems in the Artificial Intelligence Era. *British Journal of Educational Technology*, *51*(5), 1657–1673. DOI: 10.1111/bjet.12998

Wang, T., & Cheng, E. C. K. (2021). An Investigation of Barriers to Hong Kong K-12 Schools Incorporating Artificial Intelligence in Education. *Computers and Education. Artificial Intelligence*, *2*, 100031. DOI: 10.1016/j.caeai.2021.100031

Warschauer, M. (2004). *Technology and social inclusion: Rethinking the digital divide*. MIT press.

Watters, A. (2017). Dunce's App: How Silicon Valley's brand of behaviorism has entered the classroom. *The Baffler,* Weston-Sementelli, J., Allen, L. K., & McNamara, D. S. (2018). Comprehension and Writing Strategy Training Improves Performance on Content-Specific Source-Based Writing Tasks. *International Journal of Artificial Intelligence in Education*, *28*(1), 106–137. DOI: 10.1007/s40593-016-0127-7

World Bank. (2016). *World Development Report 2016: Digital Dividends.* https://www.worldbank.org/en/publication/wdr2016

World Bank. (2021). Digital economy for Africa initiative. https://www.worldbank.org/en/programs/all-africa-digital-transformation

Yang, Y., Chen, L., He, W., Sun, D., & Salas-Pilco, S. (2024). Artificial Intelligence for Enhancing Special Education for K-12: A Decade of Trends, Themes, and Global Insights (2013–2023). *International Journal of Artificial Intelligence in Education*. Advance online publication. DOI: 10.1007/s40593-024-00422-0

Yunga, F., Tapia Morquecho, C. A., Guerrero Riofrío, P. Y., & Flores Chamba, J. E. (2023). The Effect of Technology on Income Inequality: Implications of the Digital Divide — Evidence for OECD Member Countries. *Accounting and Management, 68*(1), 260–288. DOI: 10.22201/fca.24488410e.2023.3308

Zawacki-Richter, O., Marín, V. I., Bond, M., & Gouverneur, F. (2019). Systematic Review of Research on Artificial Intelligence Applications in Higher Education – Where Are the Educators? *International Journal of Educational Technology in Higher Education, 16*(1), 1–27. DOI: 10.1186/s41239-019-0171-0

Zugaza, L. A., & Ramos, J. M. L. (2007). Corruption, Development Aid, Poverty, and Human Development. *Economic Bulletin of ICE*, (2917).

# Chapter 2
# Sustainability and Its Development in Educational Practices in Primary Education in Spain

**Francisco José Martínez Carmona**
https://orcid.org/0009-0006-6224-175X
*UNIE University, Spain*

**Rubén Madrigal Cerezo**
https://orcid.org/0009-0000-5994-0564
*UNIE University, Spain*

## ABSTRACT

*Sustainability is a content of great importance to be implemented in the teaching-learning process from the early stages of the Spanish curriculum. This is indicated in its educational law, alluding to the concept of sustainability and its approach from the work of the Sustainable Development Goals (SDGs), in order to comply with the 2030 Agenda. Globally, almost 90% of people who attend primary education complete it, although the disparity in the regions where it is provided is still very uneven. This disparity has a number of causes, ranging from the availability of the necessary financial resources to be able to provide education worldwide for all girls and boys, to the lack of infrastructure in other cases. Thus, it is essential that a number of competencies that are essential for a successful life in society are acquired in primary education, and sustainability, through its Education for Sustainability (ESD), is one of them.*

DOI: 10.4018/979-8-3373-2802-7.ch002

## INTRODUCTION

Education for Sustainable Development (ESD) has been consolidated over the last two decades as an axis on which education policies are largely developed at international level and therefore in the Spanish education system. Within the framework of the 2030 Agenda, adopted by the United Nations in 2015,

Sustainable Development Goal 4 not only promotes quality, inclusive and equitable education, but also, in its target 4.7, determines that by 2030 all students should acquire the necessary knowledge to promote sustainable development, including respect for the environment, human rights, the culture of peace and global citizenship (UNESCO, 2020). This vision has been contemplated and promoted and reinforced by the institutions of the European Union, which recommends that sustainability be integrated in a cross-cutting manner at all educational levels (European Commission et al., 2024).

In the case of Spain, Organic Law 3/2020 (LOMLOE) clearly incorporates the sustainability approach, both from a cross-cutting perspective and through specific content in areas such as Civic and Ethical Values or other curricular knowledge areas. Thus, plans such as the Environmental Education Action Plan for Sustainability (PAEAS, 2021-2025) promote a transformative education oriented towards ecological transition and social justice. As Guardeño (2025) states, this educational dimension should not be considered as an add-on, but as an essential part of the curriculum, capable of articulating meaningful learning that connects the local environment with global challenges.

However, the normative presence of ESD alone does not guarantee its effective implementation in the classroom. As UNESCO (2022) points out, "environmental education cannot be reduced to a set of one-off activities, but must be structurally embedded in teacher planning and school culture". Hence the need to investigate how ESD is being put into practice at the Primary Education stage in Spain, what methodologies are used, what difficulties teachers perceive, and to what extent there are adequate resources and institutional support for its development. This paper starts from this premise and aims to analyze the degree of real integration of sustainability in everyday teaching practice, through a study of the regulatory framework, available resources and recent scientific production in the field of ESD in primary education.

In the early stages of education, and especially in primary education, many of the attitudinal and, to a large extent, cognitive foundations that will enable students to develop critically and responsibly in their social environment in the future are built (UNESCO, 2020). Thus, Education for Sustainable Development (ESD) becomes a fundamental educational approach to equip students with the tools to acquire a comprehensive understanding of the current complexity of global challenges and to act in an ethical and committed manner (European Commission et al., 2024). However, the presence of references to sustainability in curricular documents does not, in general, in itself guarantee an effective implementation in classrooms or in the development of teaching plans, nor does it ensure that teachers have the necessary didactic and organizational resources at their disposal, nor the training and support that are necessary for their real integration in the current educational context (Tilbury, 2011). In this sense, we believe it is important to review and analyze to what extent schools and educational centers are integrating ESD into their daily practice, what methodologies they are using and applying in the classroom and what difficulties or barriers teachers encounter, mainly, to apply it in a cross-cutting, real and practical way.

There has undoubtedly been normative and regulatory progress on the part of educational administrations, but significant differences can still be detected between what is proposed in legislative texts and what is actually put into practice in schools, so it seems clear that there is a need for studies that analyze this difference between theory and educational reality (Bourn, 2020). Therefore, this work arises from an understanding of the need to understand how ESD is being implemented at the Primary Education stage in Spain, while trying to identify both the real progress that is being made and the obstacles that may be hindering full curricular integration. The general objective of this study is to carry out a systematic and well-founded analysis of the incorporation of Education for Sustainable Development (ESD) in the regulatory context of the Spanish education system, in the teaching resources aimed at teachers in general and in the scientific publications that have been produced in recent years, in order to assess its degree of implementation at the Primary Education stage.

Several publications point out that the success of ESD depends not only on its presence in curricular documents, but also on its effective integration into daily teaching practice and, to a large extent, on the institutional commitment to education that can be considered transformative (Novo, 1998; Sterling, 2011). The analysis presented here also seeks to provide concrete evidence that will serve as a basis for possible future decisions on pedagogical planning and management, contributing to the design of more effective teaching strategies to achieve the integration of sustainability in real and current educational practices.

In short, this paper is intended as a contribution with the main objective of carrying out a critical analysis of the place of sustainability in the Spanish education system, and specifically in the Primary Education stage. From a global and integrating point of view, the aim is to generate as clear and real a vision as possible of the current state of ESD, not only from a regulatory or theoretical point of view, but also in terms of its practical projection. This approach will enable the identification of what has been achieved, while highlighting the remaining challenges that still need to be addressed in order to practice education that is truly committed to sustainable development.

## SUSTAINABILITY IN PRIMARY EDUCATION IN THE EUROPEAN CONTEXT

The European educational context shows us the different angles from which sustainability is approached in the classroom, marked by the guidelines of the European Union itself or of international organizations such as the United Nations Educational, Scientific and Cultural Organization (hereinafter UNESCO). The latter has developed a sustainable development plan from 2020 to 2030. The foundations were laid in the Berlin Declaration on Education for Sustainable Development signed by the member states of this organization, including most of the European states. In it, the main commitment is to integrate ESD into the different educational levels, including the primary education stages, to promote the involvement of students in a significant way in sustainable development through participation in society, recognizing the transcendental role of teachers and the use of technology in an inclusive way. Thus, the commitment starts from the international sphere in order to reach the local sphere, monitoring the aforementioned bases and plans established through the SDGs for their fulfilment, with special attention to Target 4.7 of SDG number 4 (ensure knowledge of sustainable development in students through concrete actions).

The Council of the European Union issued a series of recommendations aligned with UNESCO and its foundations, seeking the activity of the entire education community to achieve the goals set. In this way, the different European countries are adapting their educational policies for the implementation and consolidation of ESD in the different educational stages, although in this study we will focus on those concerning primary education or the first stages of the teaching-learning process. In turn, a framework of competences ("GreenComp") has been created that seek to embody the values of sustainability, embrace the complexity of sustainability, act for sustainability and envisage sustainable futures.

In common, the participation of several schools can be found through the Eco-schools program, where countries such as Denmark, Turkey, Italy, Germany or Bulgaria, among many others, follow the seven steps that form their working methodology through the formation of an Ecological Committee led by students and that can be formed by the educational community as a whole to sustainably audit their school and community to develop an action plan including curricular elements in it. Once an Eco-

Code has been assessed and developed, achieving the seven steps, a Green Flag can be applied for and obtained through an external assessment. On an individual basis, EU countries also carry out educational actions to favor such curricular integration, independently of the adherence to the Eco- schools program. We will see some cases, from here on, of how sustainability is worked on in the context of primary education in various European countries.

France establishes education for the environment and sustainable development in primary education, from the perspective of raising awareness of the environmental situation and the ecological transition. This is included in its current education law in article L312-19, which also includes other more specific points, such as the fight against climate change or the reduction of waste, the reuse of materials or their recycling. In terms of the subjects taught, there are a total of five, where there is also a reference to eco- delegates in schools and the awarding of an E3D seal, which shows the interest in ESD in that country (Guardeño, 2025).

In Italy, sustainability is addressed in the subject of Civic Education, which refers, among other things, to the protection of the environment and natural resources, the creation of inclusive living environments and respect for fundamental rights, which are included in one of the subject's core areas called Sustainable Development (Scala et al., 2024).

Following the work of other countries on ESD, we find the case of Ireland, where five of the eight competences established in the curriculum can be related to ESD and where it is taught in various subjects (Guardeño and Monsalve, 2024). The same authors point out some of its areas, such as care and respect for the environment.

Within the training established in Germany for the development of sustainability, although it has been implemented since the 1980s through programs related to environmentalism in schools and the subsequent transformation towards ESD (University of Bielefeld/Oberstufenkolleg, 1999). Today, in addition to its legislation, Germany is also strengthening its action for ESD work through the SDGs in its German Sustainability Strategy. This plan aims, in the decade in which we find ourselves, to ensure that students at all levels of education acquire competences for sustainable development, acquire training in research and science so that any decision on ESD is based on it, and that in innovation processes (particularly in the context of digitalization), sustainability aspects are taken into account to minimize risks to the environment and people. Thus, the different federal states pursue their policies within the German regulatory framework. Let us look at two examples. In the State of Bavaria, the "Schools for Life" program has been implemented in primary and secondary schools. This program is based on a five-week project (five days in two parts or continuously) related to health and nutrition education, environmental protection, consumer education including financial literacy and "foresight for life and environmentally conscious and sustainable lifestyles" (Ministry of Education and Cultural Affairs of the State of Bavaria, 2025). The other example can be found in the State of Berlin or Berlin-Brandenburg Region, where consumer education cuts across the established subjects. Thus, the same topics are dealt with as in the previous region, with the addition of media and information in this respect. Finally, students are expected to be able to recognize the ecological, economic and social consequences of individual consumption and the actions necessary to make decisions consistent with sensible consumption.

Turkey has the peculiarity that its curricular framework does not establish sustainability as a transversal aspect in its educational action, but it does work within a separate and optional subject, which is called Environmental Education and Climate Change (Eurydice, 2024). At a collaborative level, for example, an eTwinning project was carried out between prospective primary school teachers in Italy and Turkey to raise awareness of the causes and consequences of climate change within a framework of sustainable environmental education.

Another type of curricular organization around sustainability can be found in Romania, where primary and secondary schools have a "green week" to work on all kinds of related aspects. In addition, as in Turkey, there is also a separate subject called Ecological Education and Education for the Protection of the Environment. In the case of Sweden, the subject "Sustainable Society" is a compulsory subject in the curriculum. The link to the importance of the environment is incorporated in Croatia as part of the curriculum, but this time linking it to the promotion of healthy living and physical activity.

The Finnish case has two key points to focus on for ESD work. Iceland, for example, includes in its curriculum a subject dedicated to the environment and other subjects with content related to the conservation of the natural environment (González-Anleo and Barbero, 2023). This author also talks about the circular economy model that he intends to implement in order to avoid excessive consumption of resources, introducing this model into the Finnish curriculum. The nomenclature of the competence covering this area is called "participating, influencing and building a sustainable future" (Eurydice, 2024).

In Austria, a study published in 2023 reveals that the ECOLOG network, which aims to green schools and educate for sustainability in Austria, has been training in more than 700 schools, where the results show that pupils learned to deal with limited resources, where also the training was linked to new ways of thinking and teaching (Franz and Dulle, 2023). Furthermore, its curriculum implements sustainability from the primary school stage as a cross-cutting competence through environmental education for sustainable development, independently of economic, financial or consumer education, as is the case in a number of other countries (Federal Consolidated Act, 2024). It is also alluded to that the teaching- learning process should empower students to address social, economic and ecological challenges.

The Slovenian Law on the Organization and Financing of Education (ZOFVI) includes education and training for sustainable development as one of its educational objectives. Within this objective, active inclusion in a democratic society, individual responsibility, personal and collective health and culture or towards the natural and social environment are mentioned, with reference also to future generations.

Denmark and the Netherlands do not define sustainability in their curricula because schools have the autonomy to define their teaching content. More specifically, in Denmark, sustainability competences are taught on a project basis and in the Netherlands, Albania or Bosnia and Herzegovina, in separate thematic areas (Eurydice, 2024).

We can see how the different countries incorporate sustainability into their educational frameworks in one way or another, although there is still no total consensus, steps are being taken by the European Union to develop competences from ESD.

## SPANISH CURRICULUM: APPLICATION OF SUSTAINABILITY IN THE PRIMARY EDUCATION STAGE.

In the Spanish Primary Education stage, the curriculum is structured around some elements that the curriculum itself determines, which are: the key and specific competences of each subject, the basic knowledge, the assessment criteria and the learning situations which, understood in a synthetic way, are the units of the didactic planning which will be contextualized in such a way as to enable pupils to acquire the competences and learning that are relevant and significant for their daily lives, while encompassing all the curricular elements and considering the principles of UDL (Universal Design for Learning) to present an inclusive and effective design.

Sustainability, in this curricular framework, is not contemplated as a specific subject, but rather as a transversal treatment that should permeate the different areas of knowledge that are developed in the different courses of the stage, forming part of the integral growth of students (Navarro-González, I., & Gavari-Starkie, E, 2024). This curricular design promoted by LOMLOE incorporates references to sustainability explicitly in the students' exit profile, in different specific competences, linked above all to the project-based learning methodology. Thus, schools are expected to design didactic proposals that integrate environmental, social, economic and also ethical issues from a critical and participatory perspective.

This curricular approach recognizes that current challenges -such as climate change, inequality or biodiversity loss- require an education capable of promoting responsible attitudes, systemic thinking and the capacity for social transformation from an early age (Sterling, 2011; Novo, 2006). Sustainability therefore becomes a guiding principle for learning, although its concrete application depends, to a large extent, on the work of the teaching teams and the institutional commitment of each school.

In 1999, the *Libro Blanco de la Educación Ambiental en España (White Paper on Environmental Education in Spain*) was published, in which environmental education is approached as a fundamental strategy for advancing towards a model of sustainable development. This document defines the theoretical foundations, principles, objectives and priority lines of action, underlining the need to integrate the environmental perspective in all policies, especially in education. The importance of the existence of an informed and participative citizenry, together with institutional coherence, the involvement of social agents and the necessary training of teachers, is noteworthy. Already, transformative educational action is proposed, understood in an interdisciplinary and continuous manner, centered on values and with a clear orientation towards the resolution of real environmental problems (Pereira, 2003).

In the regulatory scope of LOMLOE, Royal Decree 157/2022, of 1 March, which establishes the organization and minimum teachings of Primary Education, this regulation clearly reinforces this orientation as it includes Education for Sustainable Development (ESD) in the area of Education in Civic and Ethical Values. This area focuses on knowledge of and respect for human rights, gender equality, the culture of peace and non-violence, and respect for the environment, and seeks to promote citizenship committed to current global challenges.

Different studies and official publications analyze both the Spanish regulatory framework and the inclusion of ESD in teaching practice, highlighting the importance of the involvement of the entire educational community, as well as the use of active methodologies to promote sustainability from an early age. Below is a list of some important references that address the application of sustainability in primary education in Spain from the different perspectives we present: regulations, pedagogical proposals, academic research and practical experiences in schools.

Thus, the Pedagogical Guide for learning about the Sustainable Development Goals (SDGs) in Primary Education, published by the Ministry of Social Rights, Consumption and Agenda 2030, is an official document that details how sustainability and environmental awareness are integrated transversally into the primary school curriculum, promoting the active participation of pupils while developing key competences to face the challenges of the 21st century.

SOStenibilidad, is a didactic material published by the Ministry for Ecological Transition and the Demographic Challenge (Huertas Alcalá, S. 2021). It is a didactic proposal aimed at primary and secondary education stages which includes proposals for working on sustainability in the classroom, the proposed methodologies give the students a leading role and promote educational processes focused on achieving sustainability in educational centres.

Education for Sustainability in Spain. Reflections and proposal (del Álamo et al., 2017) is a report that has been prepared by educational experts that analyses the state of Education for Sustainability in the education system and proposes lines of action and improvement, including the primary school stage.

Education for Sustainability in the minimum teaching of Primary Education Analysis of the LOMLOE (Varela-Losada et al., 2024). This is an academic article that compares the incorporation of sustainability in the educational curricula of the LOMCE (Organic Law 8/2013, of 9 December, for the improvement of educational quality) and the LOMLOE, in which advances, shortcomings and the need for adequate teacher training and community involvement are highlighted.

Teacher involvement in primary education on environmental education and sustainable development in Spain (Díaz, 2024) is a study that analyses the degree to which sustainability has been implemented in Spanish primary education classrooms. This study shows the commitment of teachers and the relevance of school projects to involve the whole educational community in sustainable education.

The Sustainable Development Goals in primary education and their implementation through ICT (Piñeiro Méndez, 2022). It shows an analysis of the integration of the SDGs in Spanish primary education, with special attention to the use of information and communication technologies as a facilitating tool to achieve education for sustainability.

CENEAM (National Centre for Environmental Education), an organization that depends on the Ministry for Ecological Transition and the Demographic Challenge, has carried out a comparative study of regional environmental education strategies (Pardellas-Santiago et al., 2024). In this report, it compares different environmental education strategies in different autonomous communities, allowing the identification of good practices and innovative approaches in the integration of sustainability in the school environment.

Studies have been published comparing the different actions taken by the regional education administrations, such as the Comparative study of the Environmental Education for Sustainability Action Plan (PAEAS) with the regional environmental education strategies (PAEAS Network, 2024). This report provides a detailed analysis of the environmental education strategies adopted by eight Autonomous Communities, comparing them with the Environmental Education for Sustainability Action Plan (PAEAS). It can be seen that most of the Autonomous Communities have designed their own strategic plans, with various actions aimed at teacher training, citizen participation, conservation of the natural environment and the application of sustainable practices in educational centers. It is worth mentioning that some Autonomous Communities, such as La Rioja, Navarra and Euskadi, stand out for integrating specific objectives that are related to action on climate change and the

SDGs of the 2030 Agenda. Good practices can be identified in promoting the development of networks of sustainable

schools, eco-schools programs and collaborative actions between public administrations. However, important differences are also highlighted, particularly with regard to the regularity of plans over time, the funding allocated and the existence of evaluation systems that provide conclusive data. The study concludes that, although progress has been made, there is a need to strengthen territorial coherence and the actual implementation of commitments.

In general, in addition to the Autonomous Communities that have developed consolidated strategies and action plans with a high level of detail, other Autonomous Communities show a less uniform degree of progress in the implementation of environmental education. Some territories, such as Castilla y León, Aragón and Extremadura, have recently approved strategies that incorporate evaluation mechanisms and multi-annual funding. Others, such as the Valencian Community, have implemented sectoral plans or programs with a significant educational impact, although without a formal strategy that is clearly defined and aligned with the SEAP. Other regions have not renewed their plans or still lack a visible strategic structure, reflecting a lack of homogeneity between the different territories that make up the country. This diversity shows that there is a need to improve coordination between the different administrations by reinforcing coordination between regional policies and the frameworks established by the state with regard to education for sustainability.

As can be seen from the articles and publications consulted, in addition to the regulatory framework established by LOMLOE, other initiatives have managed to implement environmental education in primary education. For example, Ecoembes' "Naturaliza" project seeks to integrate environmental education, albeit in a cross-cutting manner that is more common in the primary school curriculum, by providing training and teaching resources to teachers to address issues such as climate change, biodiversity and the circular economy (Carbajo Blanco, A, 2021). The programme "Centros Educativos Hacia la Sostenibilidad" in La Rioja promotes collaboration and support campaigns with schools, including workshops on domestic ecology and visits to facilities such as wastewater treatment plants, in a way that complements the teaching work in the classroom (Ministerio para la Transición Ecológica y el Reto Demográfico, n.d.). On a more academic level, Espinosa-Cevallos (2024) stresses the importance of introducing education for sustainability in school curricula in compulsory education, where it is important to integrate environmental, social and economic aspects in order to train students to face emerging environmental challenges. The denaturalization of school playgrounds is becoming a common practice, so that school spaces dominated by concrete are transformed into more inclusive and natural environments that promote diversity and contact with nature. All these experiences show that environmental education is consolidating, albeit at a slow pace in educational practice, not only as a component of the curriculum and educational programs, but also as a good tool to promote sustainability and environmental commitment from an early age.

## ONLINE RESOURCES AVAILABLE TO TEACHERS IN SPAIN FOR ESD WORK

There are currently numerous resources available on the Internet for teachers to extract different learning situations or educational experiences related to ESD. All of them can be replicated in the corresponding educational environment, taking into account the context to which they are addressed, or as a source of inspiration to carry out other types of learning situations derived from and adapted to state or regional legislation, as we have seen in previous points.

The Environmental Education for Sustainability Action Plan (hereinafter, PAEAS), developed by the Coordinating Group of the Environmental Education for Sustainability Action Plan of the Ministry for Ecological Transition and the Demographic Challenge (hereinafter, MITECO) and the MEFPD, sets out objectives and lines of action to be implemented between 2021 and 2025. More specifically, and for the area of primary education, it redirects us to various state bodies, which we will see below, to obtain a series of resources available for teachers.

The first of these is the National Centre for Environmental Education (hereinafter, CENEAM). On its website there is a section on resources, which offers multiple options for public consultation. Among the most accessible are its Digital Library, which provides 1398 documents for direct and free download or by loan request (articles, books, games and books). If we filter it by the educational stage we are dealing with, "primary", the search is limited to 201 results. If we go to the articles section, we find proposals related to fauna, sustainable mobility or the construction and use of school models. It should be noted that most of the articles date from 2009 to 2011. With regard to the books, we find publications that are much more up-to-date, with books dating from the year 2024. This is a largely up-to-date source of resources, as books are the main resources offered by the CENEAM. Biodiversity, sustainable thinking, critical education, energy eco-audits, etc., are some of the topics proposed in these books, always within the framework of ESD. Not only national proposals or publications are included, but also international proposals from countries from different continents, including America and Oceania, are available to the educational community. We can even see updated audiovisual resources that can serve as an introduction to a learning situation or create gamified experiences, among others.

The second resource is that of the Fundación-Biodiversidad, which depends on the MITECO, where all the projects carried out in the primary school stages can be consulted in its databases. Many of these projects have already been carried out, but what can serve as a basis for teachers are two aspects: to know the feasibility of whether any of these projects can be carried out in the educational center where they are working as teachers and to be able to analyze the reports collected in the results of each project, as well as the established guides, which can serve as a source of inspiration to develop proposals adapted to the more local context. In order to access this type of document, it is important to go into each project and the different items that emerge from it.

The third and final resource provided by the PAEAS is the MEFPD itself, more specifically in its Sustainable Development area linked to the Sub-Directorate General for Territorial Cooperation and Educational Innovation. Within the information available, there is a block of resource content, both of a ministerial nature and others related to other organizations. In addition to guides or manuals, as we have seen in the previous resources, there is also a very valuable resource aimed at improving the teaching-learning process. This resource is a manual of good practices, which brings together almost twenty of them. Finally, there is also a document with links to resources offered by each Autonomous Community, different institutions and organizations. This document also categorizes different aspects of ESD, such as Global Citizenship, Education in Values, Coeducation or Energy Efficiency and Renaturalisation, among others. For this reason, it has been decided not to carry out an analysis by Autonomous Community, as the resources associated with them on ESD are concentrated on the MEFPD website.

Another source that does not appear in the PAEAS but which has other types of resources for teachers is the National Institute of Educational Technologies and Teacher Training (hereinafter, INTEF). The first of these are learning situations applicable in primary education for all cycles of the stage. You can also find inspiring experiences or didactic itineraries. Finally, there is the Procomún resource search engine, which can be accessed from the INTEF website and which offers more than 60 resources related to sustainability in primary education. The search engine shows the areas of knowledge, the teaching methodology used and the type of educational context for which the available resource has been designed. It can be seen that most of the projects located coincide with active methodological work.

Finally, we summarize in table form the resources found from the PAEAS and the other sources analyzed and the type of material they provide.

*Table 1. Online material available in Spain for ESD by institution or organization.*

| Institution or organization | Material available |
|---|---|
| **National Centre for Environmental Education (CENEAM)** | Audio-visual resources, books, articles, games and stories |
| **Foundation-Biodiversity** | Projects developed, guides and manuals |
| **Ministry of Education, Vocational Training and Sports (MEFPD)** | Guides, generic manuals, good practice manuals and manuals that concentrate resources by Autonomous Community and others categorized by ESD work block. |
| **National Institute of Educational Technologies and Teacher Training (INTEF)** | Learning situations, innovative experiences, didactic itineraries and own resource search engine. |

Source. Own elaboration

## DISCUSSION AND CONCLUSIONS

By making a general assessment of the integration of ESD in the Spanish education system, we can begin to conclude, as this work shows, that although Education for Sustainable Development (ESD) is explicitly incorporated into the regulatory framework of the Spanish education system, its full integration into the curriculum and into real educational practice in the classroom still presents significant inequalities and shortcomings.

The entry into force of LOMLOE and the curricular norms derived from it is an important step forward as it includes sustainable development as one of the fundamental principles of the Spanish education system, both in its formulation and in its curricular implementation. However, in order to make the transition from legislative standards to a real transformation of school culture, institutional commitment, adequate and continuous teacher training and resources that are sustained over time (Aznar-Minguet, P et al., 2017) are needed, even though they are important, more than legal references.

The student exit profile for the primary education stage, understood as a description of the key competences, will most likely clearly define the type of critical citizenship committed to the environment that is intended to be formed. These intentions do not always have a correlate and are not always clearly reflected in didactic and teaching planning documents in a concrete way, nor on many occasions in clear assessment indicators. The cross-cutting nature of sustainability poses a real risk of dilution, precisely because it is not always explicitly linked to clearly identifiable pedagogical strategies (Vega-Marcote et al., 2015).

Looking from a strategic level, as can be seen from the articles and different studies consulted, what we could call the Spanish Strategy for Sustainable Development 2030 and the Environmental Education Action Plan for Sustainability (PAEAS) are very useful frameworks for guiding educational policies, programs and actions. However, the translation of these frameworks into real educational and concrete practice is still uneven across the different regions of the country, as can be seen from the comparative analysis carried out by the PAEAS network (2024). Some autonomous communities, such as La Rioja, Navarre and the Basque Country, have developed very specific educational plans on sustainability with monitoring, evaluation and the creation of networks of centers. On the other hand, it is observed that other autonomous communities still do not have clear and renewed strategic

frameworks, which could lead to a significant territorial gap in terms of educational opportunities in an eco-social key, . Therefore, the existence of environmental education programs such as CEHS (Educational Centres for Sustainability) or Naturaliza, which have achieved relevant success in linking curriculum, sustainability and community participation in a significant way, is highly valued. These initiatives reflect the importance that transformative education can have from a practical approach, as advocated by Sterling (2011), so that learning is not limited to the classroom, but connects with everyday life, the territory and student decision-making.

With regard to the limitations detected in the publications consulted on the implementation of ESD at the primary education stage, the one that stands out the most is related to specific teacher training, both initial and ongoing, which is considered to be very scarce. The studies analyses show that many teachers, although they consider it important to practice education for sustainability, lack methodological tools or clear references to do so in depth (Del Pozo & Novo, 2017). In the end, this often leads to ESD being treated superficially, and often restricted to specific dates or activities which, although they belong to the work planning in schools, these activities have no connection with the curriculum or evaluation processes.

Another limitation is the lack of clear indicators to assess the real impact of ESD implementation on student learning and its consolidation. Despite policy and regulatory developments, there does not seem to be a consolidated culture of formative evaluation and sustainability-oriented monitoring of ESD. This lack may hinder effective policy development, clearly limiting the capacity for continuous improvement in the development of authentic ESD education.

The rigidity of the curriculum in primary education has also been revealed as a difficulty in implementing ESD in the classroom, the pressure to meet and achieve traditional academic standards, although less than in later curricula, and the shortage of time in school timetables do not help the implementation of active methodologies that are also focused on sustainability (SauvÃ, 2005).

We consider that this work offers a vision that relates the Spanish regulatory and educational framework with real practical experiences applied by schools according to the studies consulted together with recent publications. It has become clear that

Education for Sustainable Development (ESD) should not and cannot be understood solely as a content, but must be treated from a comprehensive pedagogical perspective that affects the entire curriculum, school organization and the relationships established with the environment. The analysis carried out allows for the identification of good practices that can be replicated and guide future pedagogical and political actions. For example, the project "Every drop counts", which has been promoted by the Ministry of Education, has brought together teachers from different Spanish territories with the intention of promoting awareness of the responsible use of water and environmental sustainability through active methodologies (Núñez et al., 2018).

In order to achieve an effective integration in the teaching-learning processes currently in place, it is also clear from the analysis of the studies consulted that the need to integrate ESD in the initial teacher training processes is important, as well as in the pedagogical planning of educational centers.

Therefore, in order for education for sustainability to be consolidated in the Spanish education system, and for it to be evident in everyday teaching practice, it is necessary to move in several directions:

- Systematically incorporate ESD in the planning of teacher training, both initial and in-service.
- Establish specific evaluation indicators to measure the impact of ESD on student learning, on the development and acquisition of the key and specific competences established in the curriculum and, therefore, of eco-social competences.
- Provide schools with greater autonomy and resources, both didactic and organizational, to develop sustainable projects contextualized in the immediate environment of children, and close to their daily reality.
- Encourage the application of active teaching methodologies such as: project work, service- learning and active student participation as key methodologies.
- Promote networks of schools, communities of practice and seek synergies with agents not only in the educational sphere but also in the social and environmental sphere. Educational sustainability is not only an objective, but could become a way of understanding the educational process in a participatory, interdisciplinary and action-oriented way, while at the same time committed to ecological and social justice (Tilbury, 2011).

In conclusion, we could say that education for sustainable development represents an opportunity to rethink the deeper meaning of education in the 21st century. It should not just be about adapting content or introducing new learning situations to standard curriculum content, but about transforming the way we teach, learn and relate to the world. This work has shown that, despite normative and regulatory advances and some significant experiences, there is still a long way to go for ESD to become a hallmark of the education system, although it has also shown that there are conditions, ideas and specific experiences that allow us to affirm that this path is possible and can be achieved. The effective and practical implementation of ESD requires a real and collective commitment that fully involves the entire educational community. It is essential to foster a school culture that maintains an ethos that values sustainability, encourages active participation and generates constructive criticism of existing practices. In this way it is possible to form citizens capable of facing the challenges of the 21st century with responsibility, creativity and solidarity.

## REFERENCES

Aznar-Minguet, P., Ull, M. Á., Martínez-Agut, M. P., & Piñero, A. (2017). Evaluar para transformar: evaluación de la docencia universitaria bajo el prisma de la sostenibilidad. Enseñanza de las ciencias: revista de investigación y experiencias didácticas, 35(1), 5-27.

Bourn, D. (2020). *Understanding Global Skills for 21st Century Professions*. Palgrave Macmillan., DOI: 10.1007/978-3-030-44154-1

Carbajo Blanco, A. (2021). Ecoembes: Implementing innovation and social responsibility strategies.

del Álamo, J. B., Albero, C. M., Hidalgo, D. A., & Bastida, J. M. G. (2017). *Education for sustainability in Spain: reflections and proposals*. Madrid, Spain: Fundación alternativas.

Díaz, A. A. (2024). Implicación docente en educación Primaria sobre la Educación Ambiental. *Journal of Environmental Education and Sustainability*, *6*(1), 1203–1203.

*Environmental Education Action Plan for Sustainability. PAEAS* (n. d.). Ministry for Ecological Transition and the Demographic Challenge. https://www.miteco.gob.es/es/ceneam/plan-accion- educacion-ambiental.html

European Commission. European Executive Agency for Education and Culture, Parveva, T., Birch, P., Horváth, A., Piedrafita Tremosa, S., et al., (2024). *Learning for sustainability in Europe: building competences and support for teachers and schools*, Birch, P.(editor), Publications Office of the European Union. https://data.europa.eu/doi/10.2797/521213

Guardeño Juan, M. (2025). Evolution concept of education for sustainable development and its integration into the curriculum of Spain and France. *European Public & Social Innovation Review*, *10*, 1–19. DOI: 10.31637/epsir-2024-1335

Guardeño Juan, M., & Monsalve Lorente, L. (2024). Education for sustainable development in the curriculum of Spain and Ireland. *Universidad y Sociedad*, *16*(1), 30–44.

Huertas Alcalá, S. (2021). *SOStenibilidad, una propuesta didáctica para Educación Primaria*. Ministry of Education and Vocational Training.

Ministry for Ecological Transition and the Demographic Challenge (n.d.). *Educational Centres Towards Sustainability*.

Navarro-González, I., & Gavari-Starkie, E. (2024). Sustainable and resilient education in the LOMLOE curriculum. *Revista Española de Educación Comparada*, (45), 376–397. DOI: 10.5944/reec.45.2024.37968

Novo, M. (1998). La educación ambiental: bases éticas, conceptuales y metodologías.

Núñez, M. S., Morales, M. P., & Montañés, A. S. (2018). *Cada Gota Cuenta. Actas de los Premios de Investigación e Innovación de Educación Secundaria Obligatoria*. Bachillerato y Formación Profesional de Castilla y León.

Pardellas-Santiago, M., García-Fernández, J., Majadas-Andray, J., Fernández-Ramos, J., & Sampedro Ortega, Y. (2024). *Estudio comparativo del Plan de Acción de Educación Ambiental para la Sostenibilidad (PAEAS) con las Estrategias Autonómicas de Educación Ambiental*. CENEAM / OAPN.

Pereira, R. (2003). Libro blanco de la Educación Ambiental en España. *Ministry of the Environment. Secretaría General de Medio Ambiente*.

Piñeiro Méndez, A. (2022). Sustainable development goals in primary education and their implementation through information and communication technologies.

Sauv, Ã, L. (2005). Currents in environmental education: Mapping a complex and evolving pedagogical field. [CJEE]. *Canadian Journal of Environmental Education*, •••, 11–37.

Scala, D., Aguilar Cuesta, Á. I., & Cañizares, M. C. (2024). Integrating smart cities in Primary Education: An approach towards sustainability and urban resilience. ENSAYOS [Web link: http://www.revista.uclm.es/index.php/ensayos]. *Journal of the Faculty of Education of Albacete*, *39*(2), 127–144.

Sterling, S. (2011). Transformative learning and sustainability: Sketching the conceptual ground. *Learning and teaching in higher education, 5*(11), 17-33.

Tilbury, D. (2011). Education for sustainable development: An expert review of processes and learning.

Tilbury, D. (2011). Higher education for sustainability: a global overview of commitment and progress. *Higher education in the world, 4*(1), 18-28.

UNESCO. (2020). *Education for Sustainable Development: A roadmap*. United Nations Educational, Scientific and Cultural Organization. https://unesdoc.unesco.org/ark:/48223/pf0000374802

UNESCO. (2022). *UNESCO declares environmental education must be a core curriculum component by 2025*. Unesco.org. https://www.unesco.org/en/articles/unesco-declares-environmental-education-must-be-core-curriculum-component-2025

UNESCO. (2022). *UNESCO declares environmental education must be a core curriculum component by 2025*. Unesco.org. https://www.unesco.org/en/articles/unesco-declares-environmental-education-must-be-core-curriculum-component-2025

University of Bielefeld/Oberstufenkolleg (1999). The German experience: from ecological education to education for sustainable development. Environment and Development 15 (n4) (pp 60-65) (ISSN 0716 - 1476)

Varela-Losada, M., Lorenzo-Rial, M. A., Pérez-Rodríguez, U., & Vega-Marcote, P. (2024). Education for Sustainability in the minimum teaching of Primary Education. *Didáctica de las ciencias experimentales y sociales*, (46), 89-106.

Vega-Marcote, P., Varela-Losada, M., & Álvarez-Suárez, P. (2015). Evaluation of an educational model based on the development of sustainable competencies in basic teacher training in Spain. *Sustainability (Basel)*, *7*(3), 2603–2622. DOI: 10.3390/su7032603

# Chapter 3 The Integration of Artificial Intelligence in Physical Education, Innovation, and Sustainability in Learning Performance

**Alexandra Martin Rodriguez**
https://orcid.org/0000-0002-0087-4379
*UNIE University, Spain*

**Rubén Madrigal Cerezo**
https://orcid.org/0009-0000-5994-0564
*UNIE University, Spain*

## ABSTRACT

*This chapter explores the role of artificial intelligence (AI) in transforming Physical Education (PE), highlighting its potential to personalize learning, enhance motor skill development, and improve motivation. AI tools such as motion analysis systems, virtual trainers, and gamified applications offer real-time feedback and individualized instruction, increasing student engagement and performance. Empirical evidence shows improvements in fitness, technique accuracy, and motivational climate when using AI-driven platforms. However, challenges remain, including teacher preparedness, ethical concerns about data use, and digital inequality. The*

DOI: 10.4018/979-8-3373-2802-7.ch003

*chapter argues for a responsible and inclusive integration of AI in PE through training, infrastructure, and ethical guidelines, while emphasizing the need for further research into its long-term impact*

## INTRODUCTION

The arrival and impact of artificial intelligence (hereinafter AI) in the educational field since the emergence of ChatGPT has not left any member of the educational community indifferent. Students and teachers use it for various tasks in their respective roles, schools are creating subjects related to AI and the different governments are trying to legislate on an ethical and regulated use, which includes education as one of the points to be taken into account when establishing such legislation. We establish ChatGPT as a reference and disruption because of its revolution with respect to natural language processing, further development, creation of new related applications, etc., and because the development of AI in education dates back several decades, more specifically to the 1960s, with the development of PLATO (automated teaching) at the University of Illinois (Bitzer et al., 1965).

Physical Education (PE) is one of the subjects taught in most of the educational curricula in the compulsory education stages all over the world, and it has not gone unnoticed by the arrival of AI. Among the multiple benefits offered by the application of technology in the area of PE are the following (Martín-Rodríguez & Madrigal-Cerezo, 2025), we find AI. However, it is necessary to go deeper in this regard to detect how it can be implemented effectively within the subject, since training in digital competence provides us with many possibilities within the teaching-learning process. In short, training in digital competence will always have to be aligned with motor skills as a fundamental axis of the educational action as far as the subject of PE is concerned.

Students' acquisition of adequate digital competence is linked to their use of the IA and their mobile devices or that of their families, depending on their age and access to these devices. Exposure to screens is a factor to control and avoid among children aged 0-4 years (World Health Organization, 2019). Various associations, organizations and entities have carried out studies and proposals in this regard. It is not recommended to use technology one or two hours before going to bed, it is advisable to limit the time of use or delay access to devices with screens except for justified reasons (Asociación Española de Psiquiatría de la Infancia y la Adolescencia, 2024). Logically, throughout the development of the child and adolescent, these restrictions decrease and, as a consequence, increase the time spent using technology, autonomy and temporary exposure to screens. It should be noted that the use of mobile devices and their applications for the development of sport by the youth population has increased (Ubago Jiménez et al., 2022). That is to say, educating about it is also a teaching task and, within the field of PE, as well.

Although there is not yet a significant number of studies on the use of IA in PE, there is still no significant number of studies on the use of IA in Physical Education (Bofill et al., 2025; Wang & Wang, 2024), we can begin to interpret the relevance that it can have and begin to establish the basis for its implementation in the subject. The creation of images can help us to understand the human body and its different manifestations when it comes to knowing it. This fact is also important when executing movements correctly. Moreover, there are already publications where reconstructions of the human body are proposed to be applied to the online evaluation of sport sessions (R. Guo, 2023). Knowing how to use AI in other types of applications where autonomous learning derived from aspects worked on in the classroom can be promoted as another argument to encourage the use of leisure and free time, among others. One of these applications could be the creation of variants on a sporting modality, the calculation of passing times in different physical-sports training sessions or even generating new ideas when modifying some exercises within a strength training routine, for example. Analyses carried out in this regard point out its use in PE for the analysis of athletic performance, personalized training or health monitorin (Zhou et al., 2024). In line with other types of research, the applications are in line with a development focused on the optimization of data related to the discipline, computer vision and recognition and optimization of training and innovation and evaluation, always from the use of AI (Hu et al., 2024a).

It is also important to know how AI can be used in the personalization of learning, and more specifically of movement learning in the classroom. Derived from the response to the heterogeneity presented in the different educational stages (Monge Vera et al., 2024), AI is presented as a positive element to model the learning experience, in a unique and dynamic way (López López et al., 2023). From a more specific analysis, it has to be evaluated how it can be a beneficial element for the subject of PE. This analysis can be carried out from several positions. By attending to the teaching-learning process: helping teachers in the planning and elaboration of their didactic proposals, creating elements by the students themselves according to their interests and motivations, or associating the use of IA to the motor action. One of these elements, motivation, is key for students to develop to the maximum of their capabilities. As an example, the use of the Polar Beat application has been shown to motivate students to engage in physical activity both inside and outside the educational context (Vega-Ramírez et al., 2020).

In this case the principal objective of this chapter is to investigate the capacity of AI to transform physical education via personalized learning, data-driven performance assessment, and interactive engagement methodologies. The primary objectives encompass:

1. Comprehending AI in Physical Education: Elucidating artificial intelligence and its applications in physical education, encompassing machine learning, computer vision, and wearable technologies.
2. Improving Student Engagement and Motivation: Examining the impact of AI-driven solutions, including virtual coaches, gamified fitness programs, and adaptive learning platforms, on student motivation and involvement.
3. Customized Training and Skill Enhancement: Examining how AI can deliver instantaneous feedback, insights for injury avoidance, and tailored workout regimens based on students' biomechanics and fitness levels.
4. AI-Driven Sustainable Practices: Investigating the role of AI in fostering enduring fitness habits, minimizing resource utilization, and encouraging health-oriented activities among students.

## APPLICATIONS OF AI IN PE

As mentioned above, there are several elements that influence how AI is applied in PE, ranging from the availability of technological infrastructure and teacher training to curriculum flexibility and student digital competence. The integration of AI into PE is not merely a matter of introducing new tools, but rather a complex process that involves pedagogical adaptation, ethical considerations, and an understanding of the diverse contexts in which education takes place.

In order to observe whether engagement and learning outcome was superior using AI-driven apps, an investigation was conducted in this regard using Tuby, Sports Party, Active Arcade, and Jumpr apps on 70 undergraduate students from two random sections of Physical Education at Mindano State University. The result suggested that AI-driven apps positively affect both physical fitness and student engagement (Ompoc & Aguinaldo, 2024). Regarding decision-making dynamics, they are also a very important part of PE, whether in the tactical or technical aspect of any sport discipline, self-regulation in the execution of a task in a physical activity, the choice of an itinerary on a given route, etc. In this case, we examined, in primary education students and through the use of AI that uses the motion recognition of this type of technology in expressive activities and posture reading in Taekwondo classes, the convergent thinking skills, as well as the resolution of problems of this type (S.-B. Lee, 2024). The result of the research was positive in both variables.

Moreover, AI tools, in this case of a generative nature (hereinafter, AIG), such as Claude, Gemini or ChatGPT were also used in PE (9th grade students in South Korea) through the implementation of a health and exercise unit. During 7 sessions, the AIG was used by the students through interaction with the AIG and confrontation of the information and always with the guidance and supervision of the teachers.

Questions were also asked to obtain quality information, understanding of exercise instructions or creation of appropriate exercise programs. It is concluded that AI can be implemented in a practical way in a PE curriculum maintaining and benefiting critical thinking, while benefits are observed in individualized learning and in the generation of content (Y. Lee & Kim, 2025).

Regarding the use of AIG tools in the classroom, they are also used as a means to create material resources in the classroom. A case in point is the development of an active board game where the DALLE-2 AIG application was used to generate the game design with the help of other tools such as Genial.ly or Photoshop (Prieto Andreu, 2024). This type of application was used in the university context, but it is highly extrapolable to other types of educational stages by adapting the type of game and creation of materials to the established curriculum.

AI is even being explored as a pillar for the creation of methodological strategies in the classroom. In combination with the Internet of Things (hereafter IoT), game actions are simulated and then explained via video. Subsequently, data can be collected through IoT or other formulas to custom tailor a particular aspect of the discipline, such as physical fitness, to the individual student. In addition, tasks can also be modified according to the student's progress in order to improve their learning process. It is proposed the integration, a posteriori, of an expert system in conjunction with the teacher to guide the student in an effective mastery of the technology or to increase the student's understanding of the sport modality studied. This type of methodology was studied in n=400 and from the results it was concluded that this type of teaching has a wide potential to improve the quality and effectiveness of teaching, although due to certain time and technological limitations it was not possible to analyze in detail the problems detected in the practical teaching of PE (Yu & Mi, 2023).

AI applications have also been created that play the role of educator or fitness trainer. One of them is found in a research that developed an AI-based application coined as an Intelligent Virtual Fitness Trainer (IVFIT), which aims to provide a personalized learning experience for students. In it, different avatars can be chosen and different data such as height, age, weight, etc. can be set. Thus, the application can represent and program different physical-sports activities, from dances to self-defense exercises or training activities (Mokmin, 2020). Alpuche et al. (2024) developed a web platform with the help of AI to provide exercise or training routines with the methodologies of Kolb's learning model and the Design Thinking approach as an innovative value and, in turn, ensuring that the development is adjusted to the real needs of the user. The purpose was to create an educational software applicable to PE sessions, beyond the client's needs, to make learning more effective and the teaching process more optimized. The body recognition model (limbs) was developed with AI. The platform was tested on a small group of students (n=5), where 80% provided reflections on their performance and their own proposals for improvement, which resulted in experiential learning. Thus, they concluded that the web platform developed can improve teaching in PE and help students to perform physical conditioning activities correctly (Alpuche Ortiz et al., 2024).

Following this type of applications of using AI in PE for sports teaching, we find another proposal in the same direction. In it, an intelligent sports teaching platform with IoT architecture is established using the OpenPose algorithm, DTW and ICP for sports postures when performing different movements. Key points of movements are acquired in this case and it is also presented as an innovative proposal for improving the quality of teaching in schools and universities, as it can help teachers to correct posture of movements. Finally, the scores of the students who used the intelligent platform were 10.81 points higher than those who did not use it (Luo et al., 2024). Similarly, research has been conducted in China on PE delivered online and its association with motion capture systems. Its conclusions tell us that the economic cost can be reduced compared to other similar systems used in this study and that the use of this type of application is promising and arouses much interest among its users (Zhao & Xing, 2024).

In addition to the above, teacher training is essential in this process. Studies show that part of the teaching staff is at an introductory level (Koh, 2024). In other words, the teacher's capacity must be at an adequate level in terms of knowledge, method, application and evaluation of the theory for physical educational technology (H. S. Lee & Lee, 2021), including IA. If there is no real knowledge on how to use and teach how to use IA, the performance level of students will hardly reach the expected level. There is still a lack of training for teachers.

Finally, we reflect a summary of the type of IA or IA used based on the bibliographic study carried out.

*Table 1. IA is used in the field of physical education*

| **Title of the publication** | **Year of publication** | **Summary of the type of AI applied in the context of Physical Education** |
|---|---|---|
| Application Model for Innovative Sports Practice Teaching in Colleges Using Internet of Things and Artificial Intelligence | 2023 | Combining IoT and AI in teaching strategy. |
| Application Model for Innovative Sports Practice Teaching in Colleges Using Internet of Things and Artificial Intelligence | 2023 | In combination with the Internet of Things (IoT), game actions are simulated and then explained via video. Subsequently, data can be collected via IoT or other formulas to customize for each student based on discipline-related aspects such as physical condition. |
| Design of an AI-Enhanced Medical Research-Informed Multi-Sensor-Based Online Physical Education Motion Capture System | 2024 | PE taught online and its association with motion capture (AI) systems. |
| Enhanced Design of a Tai Chi Teaching Assistance System Integrating DTW Algorithm and SVM | 2024 | Intelligent Assistance System for teaching by integrating Dynamic Temporal Deformation algorithms and the Support Vector Machine. Increased recognition of the movement and improved quality of feedback of the executed movements. |
| Evolving Professionalism in the AI Era: Implementing Generative AI in Physical Education | 2025 | Use of Claude, Gemini and ChatGPT for content generation, personalized learning and critical thinking assistance in a health and exercise unit. |
| PETSAI: Physical Education Teaching Support with Artificial Intelligence | 2024 | Support system called PETSAI to teach movements in PE by detecting by leveraging human pose estimation (HPE) the detection of the student's skeleton in each movement and using convolutional neural networks (CNN) for classification. They also offer speech recognition to perform parallel image and voice control and a score calculation algorithm. |
| Research on the Enhancement Path of Physical Education Teaching in Colleges and Universities under Artificial Intelligence Technology | 2024 | Intelligent sports teaching platform with IoT architecture using OpenPose algorithm, DTW and ICP for sports postures when performing different movements. |
| The Effect of AI-based Physical Education Convergence Program on Convergent Thinking and Problem-solving Skills | 2024 | Study in elementary school students and through the use of AI using motion recognition of this type of technology in expressive activities and posture reading in Taekwondo classes, convergent thinking skills, as well as the resolution of problems of this type. |
| The Effectiveness of a Personalized Virtual Fitness Trainer in Teaching Physical Education by Applying the Artificial Intelligent Algorithm | 2020 | Intelligent Virtual Fitness Trainer (IVFIT), which aims to provide a personalized learning experience for students. Different avatars can be chosen and different data such as height, age, weight, etc. can be established. |

continued on following page

*Table 1. Continued*

| Title of the publication | Year of publication | Summary of the type of AI applied in the context of Physical Education |
|---|---|---|
| The Impact of Artificial Intelligence (AI) Integration on Student Engagement and Learning Outcomes of Tertiary Physical Education Students | 2024 | Investigation of the effect of an AI-powered application on the learning outcome and engagement of PE students in higher education. AI-powered applications used: Active Arcade, Tuby, Sports Party, and Jumpr. |
| Plataforma Educativa para el Acondicionamiento en Línea de Educación Física | 2024 | Web platform using AI to refine and adapt training programs. |
| Paleocontacto GT: fusionando ejercicio físico con un juego de mesa (GT Paleocontact: fusing physical exercise with a board game) | 2024 | DALLE-2 For game design generation. |

Source: Own elaboration

## AI AND MOTIVATION IN MOVEMENT LEARNING

After an analysis of the application of AI in the PE classroom, it is also necessary to see how this application influences the students. In this case we will do so by analyzing aspects such as gamification methodologies, others that use real-time feedback (some previous studies will help us to see some of their results) and success stories in which the implementation of IA has led to an improvement in motivation. It is important to take into account the different contexts of application, as well as the type of experiences provided.

In view of the above-mentioned aspects and other motivations, a study was conducted with the aim of creating basic principles for the design of pedagogical-digital strategies within the area of PE through the collaboration of participating teachers and feedback from students and teachers. A series of challenges distributed throughout the academic year were designed using a gamified narrative structured in stages and a sample of 372 students from various educational centers in the 4th year of Compulsory Secondary Education and 1st year of Baccalaureate participated. Edmodo was used as an educational social network. In this case, and given the underlying motivation, it is concluded that implementation should occur on the basis of the characteristics of this study for the PE area, with consideration of the use of mobile applications and social networks. The implementation would entail the creation of a specific virtual teaching-learning environment, comprising three notable tools: a virtual content management system, a social network with both private and open spaces, and the potential for monitoring physical activity. The integration of the gamifying narrative is emphasized as a fundamental element (Montiel-Ruiz et al., 2024).

The university students of the degree in primary education, within the subject of physical activity in the natural environment, formed the sample of a study that aimed to analyze the impact of gamification on motivation in university students. Thus, the experimental group had some familiarization lessons with the gamified resource ClassCraft. In this sense, the experimental group used the platform as a resource within the sessions including the characteristics of a gamified learning situation (roles, avatars, rewards, narrative, etc.), in line with the objectives of the subject. In this case, it is concluded that intrinsic motivation does not vary but extrinsic motivation does, as it increases, especially when we talk about external regulation in the experimental group (Ferriz-Valero et al., 2020).

Microsoft Kinect, which integrates AI and augmented reality (AR), was used to capture movements in real time, processed by algorithms and post-processed with supervised learning for neural network training. This type of AI training can provide immediate feedback on form and alignment during exercise (important to reduce the risk of injury and ensure effective training). Moreover, similar applications have been discussed in the previous section. Thus, the study was carried out on n=60 (control group and experimental group) and the level of motivation (Likert scale) and the impact on injury (by experts) were measured. Thus, in this case it is stated that the potential of these innovations yields significant results on the improvement of the learning environment in increasing motivation and reducing injuries, as well as improving the accuracy of pose estimation (Shekerbekova et al., 2025).

Body expression has also been the subject of study in terms of gamification and motivation in the classroom. In this case it happened in the context of high school students using African Dance as a reference and it is indicated that this type of strategies, even needing a greater number of studies to make a significant confirmation on the results provided, could increase the motivational climate and the perception of fun in the classroom (Real-Pérez et al., 2021).

Society is increasingly aware of the importance of health and its relationship with the practice of physical-sports activities. A study related to the field of health and IA where the effectiveness of IA-enabled digital stethoscopes in improving physical performance, psychological well-being, motivation or participation of students (n=40) was investigated, yielded different results in this regard. This instrument allowed a real monitoring. Also, this instrument facilitated real-time monitoring, a significant advancement in the field. The findings indicated a notable enhancement in physical performance among students who utilized the AI-powered stethoscope, along with a significant increase in motivation levels. Additionally, a positive influence on their psychological well-being was observed (D. Sultan, B. Omarov, A. Rakhymzhanov, A. Niyazov, M. Baikuvekov, 2024).

From another point of view, in Scotland it was studied through interviews with teachers and students how successful strategies are carried out to engage unmotivated students. Taking into account the findings and teaching practice as emotional as well as cognitive (highlighting this form of meaningful, emotional and personal learning as a key factor in teacher learning), it is suggested that the compulsion to learn is reinforced by the positive approach offered by AI (Gray et al., 2019).

Finally, we see how all the studies provided are mostly very recent and that the development of experiences and methodologies with AI still has a long way to go, in addition to needing scientific support so that classroom experiences can be adapted to each context. Nor have all the ways that PE and its different curricula and educational stages can have within the integration of AI in the classroom been explored, which implies new challenges and horizons on which to direct educational research.

## AI PERSONALIZATION PF TRAINING AND INJURY

As the digital era continues to advance, the implementation of artificial intelligence (AI) across diverse sectors is steadily increasing (David et al., 2023). Fundamental aspects of AI, such as deep learning, adaptability, and sensory input, are becoming central to this technological progression (Bodemer, 2023). By leveraging sophisticated data analysis, pattern detection, and experiential learning, AI significantly enhances decision-making and problem-solving capabilities related to biomechanical or injuries problems. Thus, AI represents a significant advancement, paving the way for improvements in areas such as injury prevention, the optimization of athletic training, and advancements in rehabilitation methodologies along with pedagogy based on movement (Kaya, 2025).

### AI and Sports Sciences

In sport sciences, given its capacity to manage and interpret extensive datasets, AI is capable of delivering meaningful insights (Krishna Azith Teja Ganti et al., n.d.) .In fact, AI enhances coaching approaches through the examination of results and measurable outputs, enabling the design of tailored and accurate development plans. These systems fine-tune workload and timing, helping to prevent excessive exertion. Moreover, intelligent systems are transforming recovery processes by continuously adjusting therapeutic routines, offering immediate responses, and adapting interventions to each person's progress (Guelmami et al., 2023). However, Huang et al. point out that, while AI offers powerful tools such as data analysis and predictive modeling, there are key implementation challenges, including technical complexity, limited funding, and data privacy concerns. They emphasize that successful integration requires gradual implementation and ongoing professional development (Huang et al., 2024).However, advancements are steadily coming to light every year. In a recent study, authors Loján Cueva et al. explored the application of artificial intelligence in diagnosing low back pain through motion capture technology (Loján Cueva, 2023) . Furthermore, physics-informed machine learning approaches have demonstrated superior accuracy in detecting head impacts using data from smart mouthguards, suggesting strong potential for concussion prevention in youth and professional contact sports (Raymond et al., 2021). Collectively, these developments underscore the transformative potential of AI in promoting safer and more personalized physical activity environments.

Along with prevention, another area of new use of AI is the biomechanical field. Authors Anton Kos and Anton Umek explore the integration of biomechanical biofeedback systems in sports and rehabilitation (Kos & Umek, 2018). When traditional biomechanical analysis methods often require specialized laboratories and time-consuming manual evaluations, in contrast, AI-powered tools—particularly those integrating computer vision, deep learning algorithms, and wearable sensors—can deliver detailed movement analysis with greater efficiency and accessibility. For instance, AI serves as an effective resource for early detection, intelligent diagnostics, and the promotion of long-term health monitoring (L. Liu et al., 2024) and when integrated with wearable technology, enables advanced data processing, adaptive learning, energy-efficient operation, and real-time data acquisition, forming the foundation for intelligent sensing platforms that support personalized health monitoring (L. Liu et al., 2024; Shajari et al., 2023). Thus, regarding physical activity, they provided a comprehensive overview of accelerating motor learning and movement efficiency, for example, through case studies in sports such as golf, skiing, and swimming, the authors illustrate the practical applications and benefits of these systems in enhancing athletic performance and rehabilitation outcomes (Kos & Umek, 2018). Also, Hammes et al. reviewed the current use of AI in elite sports, highlighting its promising role in signal processing, motion modeling, and feedback delivery—particularly through wearable sensors and computer vision (Hammes et al., 2022). These studies collectively underscore the transformative potential of artificial intelligence in advancing sports science and physical education, enhancing performance analysis, and making training and rehabilitation more precise, accessible, and data-driven.

## AI and Physical Education

Moreover, in education, AI tools may craft customized instructional routines by interpreting each learner's profile, mimic diverse athletic or pupils' conditions instantly, and deliver responsive guidance to both teachers and learners, enhancing exercise-based curricula (Kaya, 2025). In their study, Kaya and colleagues examine the application of AI technologies such as educational robots, virtual reality environments, and personalized education programs in enhancing children's physical activity and movement skills. Their findings indicate that these AI applications effectively increase physical activity levels and support movement development (Kaya, 2025). Additionally, AI offers significant advantages in monitoring student performance and providing real-time feedback (Kaya, 2025; Prakash et al., 2024). The authors conclude that AI technologies make substantial contributions to physical education and movement development in children, highlighting the importance of providing necessary training for teachers to effectively utilize these technologies (Kaya, 2025) However, to harness these benefits, educators require adequate training. The integration of AI in physical education calls for upskilling teachers in data literacy, digital tools, and instructional design. Institutions must provide ongoing support to help educators transition from traditional models to AI-enhanced practices (Hu et al., 2024b). In a related case study, Ahn and Lim (2025) explored the perceptions of PE teachers regarding the use of AI in their professional practice. Teachers identified AI as a valuable assistant, capable of acting as a virtual coach, performance evaluator, and operational manager. However, they also raised concerns about student data privacy, algorithmic transparency, and the risk of becoming overly reliant on technology without adequate training (Ahn & Lim, 2025).

## SUSTAINABLE INTEGRATION OF AI IN PHYSICAL EDUCATION

The sustainable integration of Artificial Intelligence (AI) in Physical Education (PE) extends beyond technological innovation; it involves aligning digital transformation with long-term educational, social, economic, and environmental goals. This approach emphasizes not only the implementation of advanced tools but also the reconfiguration of pedagogical practices, school infrastructures, and institutional policies to promote responsible and equitable use.

### Reducing Material Use Through Digital Platforms

Traditional PE instruction often relies on significant material consumption, including paper-based assessments, printed schedules, activity logs, and physical resources such as testing equipment. The integration of AI-driven digital platforms can significantly reduce these demands. Goel et al. identifies both the opportunities and threats associated with digital transformation. It emphasizes the need for a balanced approach that maximizes the benefits of digital technologies while mitigating their environmental drawbacks. The authors advocate for strategies such as adopting energy-efficient practices, integrating renewable energy sources, and implementing circular economy principles to ensure that digital progress aligns with global sustainability goals (Goel et al., 2024). Online platforms and mobile applications powered by AI allow for automated tracking, performance analysis, and progress reporting, eliminating the need for paper-based documentation.

For example, tools such as *ClassDojo*, *Seesaw*, or AI-augmented learning management systems (LMS) can be adapted to physical education to manage attendance, performance feedback, and peer assessments. These platforms also encourage students to reflect on their learning and receive individualized feedback—functions traditionally managed through printed forms or manual teacher input (Benzizoune, 2024).In addition, AI-based movement analysis tools, such as *OpenPose*, *Kinovea*, and mobile apps like *Coach's Eye*, enable educators to assess technique and posture without the need for elaborate and consumable testing materials. This digital shift contributes to reducing waste and energy consumption associated with printing, transporting, and discarding physical educational materials (Pueo et al., 2020; Zsarnoczky-Dulhazi et al., 2024). Shreds of evidence, such as Oladele et al. illustrate in their study on green management practices in higher education, environmentally conscious policies—such as energy conservation, sustainable waste management, and the use of digital resources—positively impact institutional performance and operational cost-effectiveness (Oladele et al., 2023). Similarly, the work by Hasanova and Mammadova emphasizes the role of educational institutions as agents of sustainable development, advocating for the integration of green principles into infrastructure, curricula, and administrative procedures to cultivate a culture of environmental responsibility among staff and students (Hasanova & Mammadova, 2024).

Another notable example is the Smart School Sports Initiative in South Korea, where AI-based physical education programs use digital avatars, virtual exercise environments, and real-time feedback to reduce equipment use and increase engagement (Weis et al., 2022). The program led to a measurable decrease in the consumption of physical materials and promoted greater physical activity in both urban and rural settings (Khan et al., 2023).

In Europe, the GreenEdTech Project, funded by the Erasmus+ program, demonstrated that combining digital education platforms with wearable fitness trackers not only reduced resource usage but also streamlined teacher workloads. Teachers could track student progress remotely, plan sessions based on real-time performance analytics, and offer feedback with minimal physical materials (*Green Erasmus Project*, n.d.; Haraty & Bitar, 2019). A pilot Program involved the deployment of AI-enhanced biomechanics analysis tools in secondary schools. The program resulted in a 30% reduction in traditional testing equipment and improved assessment consistency across diverse student populations, especially those with special education needs (*AI Pilot to Help Tackle WA Teacher Workloads | Ministers' Media Centre*, n.d.; Mou, 2025).

## Promoting Sustainable Healthy Habits with Smart Tracking Tools

Sustainability in PE is not limited to environmental considerations but also includes fostering long-term healthy behaviors among students. AI-based wearable devices and tracking systems offer real-time insights into physical activity levels, sleep patterns, heart rate variability, and more—enabling students to better understand and manage their own health (GENÇ, 2023; Huhn et al., 2022). The research suggests that ChatGPT can assist in creating personalized training programs, analyzing athletic performance, and providing interactive learning experiences. By leveraging AI, educators and coaches can enhance engagement, tailor instruction to individual needs, and foster a more dynamic learning environment. However, the study also cautions about potential challenges, including over-reliance on technology and the need for proper training to effectively integrate AI tools into educational settings (GENÇ, 2023)

Additionally, smartwatches, fitness bands, and AI-integrated apps such as *Garmin Coach*, *Whoop*, or *Fitbit Premium* promote awareness of daily movement and recovery, encouraging a culture of consistent, self-managed physical activity. These tools offer personalized feedback and motivation tailored to individual goals, which has been shown to increase adherence to physical activity routines among adolescents (Kaya, 2025). These tools not only collect physiological data such as heart rate and stress levels, but also apply AI algorithms to identify patterns, detect anomalies, and provide timely interventions that support mental and physical health in academic environments. Vistorte et al. (2024), wearable AI devices have shown strong potential in educational settings by enabling real-time monitoring of students' physical and emotional states (Vistorte et al., 2024). Complementing this perspective, the systematic review by Dehbozorgi et al. 2025 emphasizes that AI-enhanced wearables are particularly valuable for mental health applications (Dehbozorgi et al., 2025). They enable continuous, passive tracking of emotional well-being, which can lead to early detection of issues like anxiety or depression, especially in youth populations. Both studies highlight the importance of user-friendly, non-invasive, and ethical design principles, stressing the need for further interdisciplinary research and institutional readiness to support the integration of such technologies in schools and universities (Dehbozorgi et al., 2025; Z. Liu et al., 2022; Vistorte et al., 2024)

By nurturing self-regulated habits through AI, educators can instill values that promote lifelong well-being, reducing future reliance on healthcare services and fostering a generation with a more proactive approach to health management—an indirect but meaningful contribution to sustainable public health.

## Preparing Schools and Educators for AI Implementation

A critical factor in the successful and sustainable integration of AI in PE lies in the readiness of educational institutions and their staff (S. BELLEZA et al., 2021). Educators often report limited exposure to AI tools and a lack of formal training in digital literacy. Without proper infrastructure and upskilling programs, the introduction of AI can exacerbate existing digital divides and create inconsistencies in instructional quality. However, Zha et al. exposed in their recent study that the AI integration in primary PE highlights the need for a holistic approach beyond mere technological advancement (Zha et al., 2025b).

To address this, teacher training must emphasize both technical competencies (e.g., using AI-driven applications, interpreting data) and pedagogical strategies (e.g., integrating AI feedback into lesson planning). Additionally, it has highlighted the importance of equipping teachers with knowledge of digital ethics, particularly in areas such as data privacy, algorithm transparency, and student safety (Schiff, 2022). Also, it must be considered, on an infrastructural level, that schools should invest in reliable internet access, compatible devices, and technical support services. While these requirements may pose initial challenges, governments and education stakeholders can adopt scalable frameworks such as the UNESCO AI Competency Framework for Educators, ensuring alignment with sustainable development goals (SDGs), especially SDG 4 (quality education) and SDG 12 (responsible consumption and production.

To address the complex challenges posed by the integration of AI in education, it is essential to adopt a comprehensive and proactive approach to teacher preparation and school readiness. Teacher training must emphasize not only technical competencies—such as the ability to use AI-driven applications, interpret learning analytics, and manage classroom technologies—but also robust pedagogical strategies that enable educators to meaningfully integrate AI feedback into curriculum planning and student assessment. This dual focus ensures that AI is not merely a tool for automation, but a complement to human-led teaching practices that enhance learning outcomes (Zha et al., 2025b).

## Cost-Effectiveness and Scalability of AI Tool in PE Programs

A persistent concern surrounding the implementation of AI in school-based PE is the cost of adoption. However, emerging evidence indicates that, while upfront investments in AI-compatible devices, software platforms, and digital infrastructure may seem substantial, these technologies often yield long-term financial benefits by optimizing resource use, enhancing administrative efficiency, and reducing teacher workload (Woodruff et al., 2023). AI-based systems can automate time-consuming tasks—such as attendance tracking, performance assessments, or individualized feedback—freeing up educators to focus more on instruction and student engagement.

Scalability is another key advantage of AI in PE. Cloud-based systems allow for centralized data management, enabling school networks or districts to standardize performance metrics, monitor progress in real-time, and streamline reporting processes across multiple sites. Gorver and Nandal emphasized that AI-powered platforms often incorporate adaptive algorithms capable of personalizing learning content to student needs, regardless of geographic or demographic disparities (Grover & Nandal, 2024). These scalable models also support integration with existing learning management systems and facilitate multilingual or accessibility features, fostering inclusive participation among diverse student populations.

In addition to reducing reliance on printed materials and physical forms of assessment, AI tools also decrease the need for in-person professional development and external consulting, further reducing long-term institutional costs. Some studies report improved return on investment (ROI) as schools see gains in student motivation, academic performance, and digital competence as a result of AI-supported instruction. These economic and educational benefits make AI a viable long-term solution for enhancing the efficiency and impact of PE programs (Eltahir & Babiker, 2024).

Therefore, rather than viewing AI through the lens of immediate cost, decision-makers should evaluate its cost-effectiveness and scalability within a broader vision of innovation-driven, inclusive, and data-informed physical education.

## CHALLENGES AND ETHICAL CONSIDERATIONS

AI systems in education often rely on collecting and analyzing large amounts of student data, including personal information and performance metrics. This raises concerns about how data is stored, who has access to it, and how it is used. Unauthorized access or misuse of this data can lead to breaches of confidentiality and privacy. For instance, AI-driven assessments involve extensive data collection, which, if not properly managed, can compromise student privacy (Schubert, n.d.)

To mitigate these risks, institutions must adhere to strict data protection regulations, such as the General Data Protection Regulation (GDPR), and implement robust security measures. Clear policies on data usage and consent should be established to maintain transparency and trust among stakeholders (The Impact of the General Data Protection Regulation (GDPR) on Artificial Intelligence, 2020.)

## Inequalities AI Access

The digital divide remains a significant barrier to the equitable implementation of AI in education. Students from low-income families or rural areas often lack access to necessary technology and reliable internet connectivity, limiting their ability to benefit from AI-driven educational tools (*AI and the Digital Divide: Bridging the Gap for Equitable Education*, n.d.; Lourdu Vesna, 2025). This disparity can exacerbate existing educational inequalities, as students without access to technology are at a disadvantage compared to their peers. To address this issue, coordinated efforts among governments, businesses, and international organizations are necessary. Investments in digital infrastructure, harmonized regulations, and initiatives to improve digital literacy are critical to bridging the digital divide and ensuring that all students have equal opportunities to benefit from AI in education (Meyer et al., 2023).

## Ethical Considerations

The use of AI algorithms in student performance evaluation introduces ethical concerns related to bias, transparency, and accountability. AI systems can inadvertently perpetuate existing biases present in their training data, leading to discriminatory outcomes (Chen, 2023). For example, AI models trained on unbalanced data may exhibit biases, resulting in disparities in scoring between different student groups. Concretely, Integrating AI into education necessitates a comprehensive understanding of Fairness, Accountability, Transparency, and Ethics (FATE) to ensure responsible and equitable implementation. A systematic review by Memarian et al. (2023) emphasizes that addressing FATE principles is crucial for mitigating biases and ensuring that AI applications in higher education are aligned with ethical standards. The study highlights the importance of developing clear guidelines and frameworks that prioritize ethical considerations, thereby fostering trust and inclusivity in AI-driven educational environments. By embedding FATE principles into the design and deployment of AI tools, educators and policymakers can better navigate the ethical complexities associated with AI integration in education (Memarian & Doleck, 2023).

Furthermore, the decision-making processes of AI systems are often opaque, making it difficult for educators and students to understand how assessments are determined. This lack of transparency can undermine trust and make it challenging to hold AI systems accountable for their decisions (Al-Zahrani, 2024; Latif et al., 2025). To ensure fairness and accuracy in AI-driven assessments, developers should use diverse and representative datasets, implement regular audits to detect and correct biases, and provide clear explanations of how AI systems make decisions. Engaging educators and students in the development and evaluation of AI tools can also help ensure that these systems align with educational values and goals. A systematic review by Ferrara (2024) highlights that biases in AI systems can originate from various sources, including data collection, algorithm design, and human interaction. As mentioned, these biases can lead to discriminatory outcomes, particularly when AI systems are used in decision-making processes that affect individuals' lives. The study emphasizes the importance of developing clear guidelines and frameworks that prioritize ethical considerations, thereby fostering trust and inclusivity in AI-driven educational environments. By embedding fairness and bias mitigation strategies into the design and deployment of AI tools, educators and policymakers can better navigate the ethical complexities associated with AI integration in education (Tavares & Ferrara, 2023).

## CONCLUSIONS AND FUTURE PERSPECTIVES

The integration of AI into physical education enhances personalized learning, precise performance assessment, motivation, and sustainable practices. AI tools such as virtual coaches, motion analysis, and gamified platforms adapt to individual needs and increase student engagement, while also supporting inclusive and eco-friendly approaches. However, successful implementation requires teacher training, ethical safeguards, and efforts to bridge the digital divide. A balanced, well-regulated use of AI—supported by research and policy—can enrich learning and promote the holistic development of all students.

## REFERENCES


Ahn, D., & Lim, H. (2025). Exploring K-12 Physical Education Teachers' Perspectives on Opportunities and Challenges of AI Integration Through Ideation Workshops. *CHI Conference on Human Factors in Computing Systems (CHI '25),* April 26-May 1, 2025, Yokohama, Japan, *1*. DOI: 10.1145/3706598.3713646

Al-Zahrani, A. M. (2024). Unveiling the shadows: Beyond the hype of AI in education. *Heliyon*, *10*(9), e30696. DOI: 10.1016/j.heliyon.2024.e30696 PMID: 38737255

Alpuche Ortiz, C. M., González Torres, J., & Corona Ferreira, A. (2024). Plataforma Educativa para el Acondicionamiento en Línea de Educación Física. *Ciencia Latina Revista Científica Multidisciplinar*, *8*(4), 13643–13656. DOI: 10.37811/cl_rcm.v8i4.13868

Asociación Española de Psiquiatría de la Infancia y la Adolescencia. (2024). *Recomendaciones de Uso de Nuevas Tecnologías en la Infancia y Adolescencia.*

Benzizoune, O. (2024). Enhancing Classroom Management and Student Engagement: The Role of ClassDojo and Gamification in Education. *Journal of English Language Teaching and Applied Linguistics*, *6*(3), 106–114. DOI: 10.32996/jeltal.2024.6.3.13

Bitzer, D. L., Lyman, E. R., & Easley, A. J. (1965). *The Uses Of Plato: A Computer Controlled Teaching System.* DOI: 10.21236/AD0623526

Bodemer, O. (2023). *Enhancing Individual Sports Training through Artificial Intelligence: A Comprehensive Review*. DOI: 10.36227/techrxiv.24005916

Bofill, J., Pla-Campas, G., & Sebastiani, E. M. (2025). Is Artificial Intelligence an educational resource in Physical Education? A Systematic Review. *Apunts. Educacion Fisica y Deportes, 160*, 1–9. https://doi.org/.(2025/2).160.01DOI: 10.5672/APUNTS.2014-0983.ES

Chen, Z. (2023). Ethics and discrimination in artificial intelligence-enabled recruitment practices. *Humanities and Social Sciences Communications 2023 10:1, 10*(1), 1–12. DOI: 10.1057/s41599-023-02079-x

David, S., Bačić, B., Richter, C., & Mundt, M. (2023). Editorial: Artificial intelligence to enhance biomechanical modelling. *Frontiers in Sports and Active Living*, *5*, 1188035. DOI: 10.3389/fspor.2023.1188035 PMID: 37188071

Dehbozorgi, R., Zangeneh, S., Khooshab, E., Nia, D. H., Hanif, H. R., Samian, P., Yousefi, M., Hashemi, F. H., Vakili, M., Jamalimoghadam, N., & Lohrasebi, F. (2025). The application of artificial intelligence in the field of mental health: A systematic review. *BMC Psychiatry*, *25*(1), 1–20. DOI: 10.1186/s12888-025-06483-2 PMID: 39953464

Eltahir, M. E., & Babiker, F. M. E. (2024). The Influence of Artificial Intelligence Tools on Student Performance in e-Learning Environments: Case Study. *Electronic Journal of e-Learning*, *22*(9), 91–110. DOI: 10.34190/ejel.22.9.3639

Ferriz-Valero, A., Østerlie, O., García Martínez, S., & García-Jaén, M. (2020). Gamification in Physical Education: Evaluation of Impact on Motivation and Academic Performance within Higher Education. *International Journal of Environmental Research and Public Health*, *17*(12), 4465. DOI: 10.3390/ijerph17124465 PMID: 32575919

GENÇ, N. (2023). Artificial Intelligence in Physical Education and Sports: New Horizons with ChatGPT. *Akdeniz Spor Bilimleri Dergisi*. DOI: 10.38021/asbid.1291604

Goel, A., Masurkar, S., & Pathade, G. R. (2024). An Overview of Digital Transformation and Environmental Sustainability: Threats, Opportunities, and Solutions. *Sustainability 2024, Vol. 16, Page 11079*, *16*(24), 11079. DOI: 10.3390/su162411079

Gray, S., Treacy, J., & Hall, E. T. (2019). Re-engaging disengaged pupils in physical education: An appreciative inquiry perspective. *Sport Education and Society*, *24*(3), 241–255. DOI: 10.1080/13573322.2017.1374942

*Green Erasmus Project*. (n.d.). Retrieved April 13, 2025, from https://project.greenerasmus.org/

Grover, V., & Nandal, M. (2024). Education system using cloud computing: A proposed model. *Educational Perspectives on Digital Technologies in Modeling and Management*, 178–194. DOI: 10.4018/979-8-3693-2314-4.ch009

Guelmami, N., Fekih-Romdhane, F., Mechraoui, O., & Bragazzi, N. L. (2023). Injury Prevention, Optimized Training and Rehabilitation: How Is AI Reshaping the Field of Sports Medicine. *New Asian Journal of Medicine*, *30–34*, 30–34. Advance online publication. DOI: 10.61838/kman.najm.1.1.6

Guo, R. (2023). Analysis of Artificial Intelligence Technology and Its Application in Improving the Effectiveness of Physical Education Teaching. *International Journal of Web-Based Learning and Teaching Technologies*, *19*(1), 1–15. DOI: 10.4018/IJWLTT.335115

Hammes, F., Hagg, A., Asteroth, A., & Link, D. (2022). Artificial Intelligence in Elite Sports—A Narrative Review of Success Stories and Challenges. *Frontiers in Sports and Active Living*, *4*, 861466. DOI: 10.3389/fspor.2022.861466 PMID: 35899138

Haraty, R. A., & Bitar, G. (2019). Associating learning technology to sustain the environment through green mobile applications. *Heliyon*, *5*(1), e01141. Advance online publication. DOI: 10.1016/j.heliyon.2019.e01141 PMID: 30705989

Hasanova, G. I., & Mammadova, Z. I. (2024). GREEN SCHOOLS: A BRIEF REVIEW. *GREEN ECONOMICS*, *2*(2), 178–183. DOI: 10.62476/ge22.178

Hu, Z., Liu, Z., & Su, Y. (2024a). AI-Driven Smart Transformation in Physical Education: Current Trends and Future Research Directions. *Applied Sciences (Basel, Switzerland)*, *14*(22), 10616. DOI: 10.3390/app142210616

Hu, Z., Liu, Z., & Su, Y. (2024b). AI-Driven Smart Transformation in Physical Education: Current Trends and Future Research Directions. *Applied Sciences 2024, Vol. 14, Page 10616, 14*(22), 10616. DOI: 10.3390/app142210616

Huang, Z., Wang, W., Jia, Z., Wang, Z., Huang, Z., Wang, W., Jia, Z., & Wang, Z. (2024). Exploring the Integration of Artificial Intelligence in Sports Coaching: Enhancing Training Efficiency, Injury Prevention, and Overcoming Implementation Barriers. *Journal of Computer and Communications*, *12*(12), 201–217. DOI: 10.4236/jcc.2024.1212012

Huhn, S., Axt, M., Gunga, H. C., Maggioni, M. A., Munga, S., Obor, D., Sié, A., Boudo, V., Bunker, A., Sauerborn, R., Bärnighausen, T., & Barteit, S. (2022). The Impact of Wearable Technologies in Health Research: Scoping Review. *JMIR mHealth and uHealth*, *10*(1), e34384. Advance online publication. DOI: 10.2196/34384 PMID: 35076409

Kaya, F. (2025). The Use of Artificial Intelligence in Physical Education and Movement Development in Children. *Spor Eğitim Dergisi*, *9*(1), 1–13. DOI: 10.55238/seder.1583190

Khan, M. K., Farouk, A., Yu, H., & Mi, Y. (2023). Application Model for Innovative Sports Practice Teaching in Colleges Using Internet of Things and Artificial Intelligence. *Electronics 2023, Vol. 12, Page 874, 12*(4), 874. DOI: 10.3390/electronics12040874

Koh, J. K. (2024). Analyzing PE Teachers' Interest and Execution in Using AI in Education. *Journal of Research in Curriculum Instruction*, *28*(1), 44–54.

Kos, A., & Umek, A. (2018). *Biomechanical Biofeedback Systems and Applications*. DOI: 10.1007/978-3-319-91349-0

Krishna Azith Teja Ganti, V., Pandugula, C., Naga Subhash Polineni, T., & Mallesham, G. (n.d.). Transforming Sports Medicine with Deep Learning and Generative AI. *Personalized Rehabilitation Protocols and Injury Prevention Strategies for Professional Athletes.*, *22*(1), 2023–3868.

Latif, E., Zhai, X., & Liu, L. (2025). *AI Gender Bias, Disparities, and Fairness: Does Training Data Matter?* Lee, H. S., & Lee, J. (2021). Applying Artificial Intelligence in Physical Education and Future Perspectives. *Sustainability (Basel)*, *13*(1), 351. DOI: 10.3390/su13010351

Lee, S.-B. (2024). The Effect of AI-based Physical Education Convergence Program on Convergent Thinking and Problem-solving Skills. *Journal of the Korean Society for Wellness*, *19*(4), 295–300. DOI: 10.21097/ksw.2024.11.19.4.295

Lee, Y., & Kim, D. (2025). Evolving Professionalism in the AI Era: Implementing Generative AI in Physical Education. *Journal of Physical Education, Recreation & Dance*, *96*(3), 8–13. DOI: 10.1080/07303084.2024.2444586

Liu, L., Pu, Y., Fan, J., Yan, Y., Liu, W., Luo, K., Wang, Y., Zhao, G., Chen, T., Puiu, P. D., & Huang, H. (2024). Wearable Sensors, Data Processing, and Artificial Intelligence in Pregnancy Monitoring: A Review. *Sensors 2024, Vol. 24, Page 6426*, *24*(19), 6426. DOI: 10.3390/s24196426

Liu, Z., Ren, Y., Kong, X., & Liu, S. (2022). Learning Analytics Based on Wearable Devices: A Systematic Literature Review From 2011 to 2021. *Journal of Educational Computing Research*, *60*(6), 1514–1557. DOI: 10.1177/07356331211064780

Loján Cueva, A. B. (2023). *Artificial Intelligence to diagnose low back pain using motion captures*. https://repositorio.yachaytech.edu.ec/handle/123456789/667

López López, H. L., Rivera Escalera, A., & Cruz García, C. R. (2023). PERSONALIZACIÓN DEL APRENDIZAJE CON INTELIGENCIA ARTIFICIAL EN LA EDUCACIÓN SUPERIOR. *Revista Digital de Tecnologías Informáticas y Sistemas*, *7*(1), 123–128. DOI: 10.61530/redtis.vol7.n1.2023.165.123-128

Luo, Y., Luo, L., & Luo, D. (2024). Research on the Enhancement Path of Physical Education Teaching in Colleges and Universities under Artificial Intelligence Technology. *Applied Mathematics and Nonlinear Sciences*, *9*(1), 20241770. Advance online publication. DOI: 10.2478/amns-2024-1770

Martín-Rodríguez, A., & Madrigal-Cerezo, R. (2025). Technology-Enhanced Pedagogy in Physical Education: Bridging Engagement, Learning, and Lifelong Activity. *Education Sciences*, *15*(4), 409. DOI: 10.3390/educsci15040409

Memarian, B., & Doleck, T. (2023). Fairness, Accountability, Transparency, and Ethics (FATE) in Artificial Intelligence (AI) and higher education: A systematic review. *Computers and Education: Artificial Intelligence*, *5*, 100152. DOI: 10.1016/j.caeai.2023.100152

Meyer, K. E., Li, J., Brouthers, K. D., & Jean, R. J. (2023). International business in the digital age: Global strategies in a world of national institutions. *Journal of International Business Studies*, *54*(4), 577–598. DOI: 10.1057/s41267-023-00618-x PMID: 37305175

Mokmin, N. A. M. (2020). The Effectiveness of a Personalized Virtual Fitness Trainer in Teaching Physical Education by Applying the Artificial Intelligent Algorithm. *International Journal of Human Movement and Sports Sciences*, *8*(5), 258–264. DOI: 10.13189/saj.2020.080514

Monge Vera, M. M., Villamagua Jiménez, G. M., Aroca Izurieta, C. E., Chico Guzmán, B. A., & López Velasco, J. E. (2024). Personalización del proceso de aprendizaje mediante inteligencia artificial. *LATAM Revista Latinoamericana de Ciencias Sociales y Humanidades*, *5*(3). Advance online publication. DOI: 10.56712/latam.v5i3.2076

Montiel-Ruiz, F. J., Calderón, A., Sánchez-Vera, M. del M., & Solano-Fernández, I. M. (2024). Redes sociales y aprendizaje móvil en Educación Física: Creación colaborativa de principios de diseño. *Edutec. Revista Electrónica de Tecnología Educativa*, *89*(89), 190–209. DOI: 10.21556/edutec.2024.89.3191

Mou, S. (2025). The integration of mobile technology in biomechanics education: Advancing knowledge and practice. *Molecular & Cellular Biomechanics*, *22*(3), 1410. DOI: 10.62617/mcb1410

Oladele, T., Oyenuga, M., Oyetunde, O. T., Oyedele, O. M., & Jaccinta, A. G. (2023). *Green Management Practices and Organisational Efficiency in a Higher Institution. Article in Budapest International Research and Critics in Linguistics and Education (BirLE)*. Journal., DOI: 10.33258/birle.v6i3.7716

Ompoc, H. M., & Aguinaldo, J. (2024). The Impact of Artificial Intelligence (AI) Integration on Student Engagement and Learning Outcomes of Tertiary Physical Education Students. *Pedagogy Review: An International Journal of Educational Theories. Approaches and Strategies*, *4*(1), 80–86. DOI: 10.62718/vmca.pr-ijetas.4.1.SC-1224-006

Prakash, J., Swathiramya, R., Balambigai, G., Menaha, R., & Abhirami, J. S. (2024). AI-Driven Real-Time Feedback System for Enhanced Student Support: Leveraging Sentiment Analysis and Machine Learning Algorithms. *International Journal of Computational and Experimental Science and Engineering*, *10*(4), 1567–1574. DOI: 10.22399/ijcesen.780

Prieto Andreu, J. M. (2024). Paleocontacto GT: fusionando ejercicio físico con un juego de mesa (GT Paleocontact: fusing physical exercise with a board game). *Retos*, *59*, 892–902. DOI: 10.47197/retos.v59.107734

Pueo, B., Penichet-Tomas, A., & Jimenez-Olmedo, J. M. (2020). Validity, reliability and usefulness of smartphone and kinovea motion analysis software for direct measurement of vertical jump height. *Physiology & Behavior*, *227*, 113144. Advance online publication. DOI: 10.1016/j.physbeh.2020.113144 PMID: 32822710

Raymond, S. J., Cecchi, N. J., Alizadeh, H. V., Callan, A. A., Rice, E., Liu, Y., Zhou, Z., Zeineh, M., & Camarillo, D. B. (2021). Physics-informed machine learning improves detection of head impacts. *Annals of Biomedical Engineering*, *50*(11), 1534–1545. DOI: 10.1007/s10439-022-02911-6 PMID: 35303171

Real-Pérez, M., Sánchez-Oliva, D., & Moledo, C. P. (2021). Proyecto África "La Leyenda de Faro": Efectos de una metodología basada en la gamificación sobre la motivación situacional respecto al contenido de expresión corporal en Educación Secundaria (Africa Project "La Leyenda de Faro": Effects of a methodology. *Retos*, *42*, 567–574. DOI: 10.47197/retos.v42i0.86124

S. BELLEZA, S., OMPOC, R., & VESTIL, C. (2021). Challenges of PE Teachers to Educational Technologies in New Normal: The Changing Role of Teachers. *International Journal of Research Publications*, *86*(1). Advance online publication. DOI: 10.47119/IJRP1008611020212320

Schiff, D. (2022). Education for AI, not AI for Education: The Role of Education and Ethics in National AI Policy Strategies. *International Journal of Artificial Intelligence in Education*, *32*(3), 527–563. DOI: 10.1007/s40593-021-00270-2

Schubert, C. (n.d.). *Research Guides: Artificial Intelligence (AI) in Education: AI and Ethics*. Retrieved April 16, 2025, from https://guides.lib.jmu.edu/AI-in-education/ethics

Shajari, S., Kuruvinashetti, K., Komeili, A., & Sundararaj, U. (2023). The Emergence of AI-Based Wearable Sensors for Digital Health Technology: A Review. *Sensors 2023, Vol. 23, Page 9498, 23*(23), 9498. DOI: 10.3390/s23239498

Shekerbekova, S., Kamalova, G., Iskakova, M., Aldabergenova, A., Abdykerimova, E., & Shetiyeva, K. (2025). Exploring the use of Artificial Intelligence and Augmented Reality tools to improve interactivity in Physical Education teaching and training methods. *Retos*, *66*, 935–949. DOI: 10.47197/retos.v66.113540

Sultan, D., Omarov, B., Rakhymzhanov, A., Niyazov, A., & Baikuvekov, M., B. O. (2024). Development of an artificial intelligence-enabled non-invasive digital stethoscope for monitoring the heart condition of athletes in real-time. *Retos: Nuevas Tendencias En Educación Física. Deporte y Recreación*, *60*, 1169–1180.

Tavares, S., & Ferrara, E. (2023). Fairness and Bias in Artificial Intelligence: A Brief Survey of Sources, Impacts, and Mitigation Strategies. *Sci 2024, Vol. 6, Page 3, 6*(1), 3. DOI: 10.3390/sci6010003

Ubago Jiménez, J. L., Puertas Molero, P., González Valero, G., Melguizo Ibáñez, E., Valverde Janer, M., & Ortega Caballero, M. (2022). Uso de los dispositivos móviles (Mhealth) en la práctica deportiva en adolescentes. *Revista INFAD de Psicología. International Journal of Developmental and Educational Psychology.*, *1*(1), 165–170. DOI: 10.17060/ijodaep.2022.n1.v1.2340

Vega-Ramírez, L., Notario, R. O., & Ávalos-Ramos, M. A. (2020). The Relevance of Mobile Applications in the Learning of Physical Education. *Education Sciences*, *10*(11), 329. DOI: 10.3390/educsci10110329

Vesna, L. (2025). Digital Divide in AI-Powered Education: Challenges and Solutions for Equitable Learning. *Journal of Information Systems Engineering & Management*, *10*(21s), 300–308. DOI: 10.52783/jisem.v10i21s.3327

Vistorte, A. O. R., Deroncele-Acosta, A., Ayala, J. L. M., Barrasa, A., López-Granero, C., & Martí-González, M. (2024). Integrating artificial intelligence to assess emotions in learning environments: A systematic literature review. *Frontiers in Psychology*, *15*, 1387089. DOI: 10.3389/fpsyg.2024.1387089 PMID: 38966729

Wang, Y., & Wang, X. (2024). Artificial intelligence in physical education: Comprehensive review and future teacher training strategies. *Frontiers in Public Health*, *12*, 1484848. Advance online publication. DOI: 10.3389/fpubh.2024.1484848 PMID: 39583072

Weis, T., Zhao, Y., Tsourdos, A., & Yu, J.-E. (2022). Exploration of Educational Possibilities by Four Metaverse Types in Physical Education. *Technologies 2022, Vol. 10, Page 104, 10*(5), 104. DOI: 10.3390/technologies10050104

Woodruff, K., Hutson, J., & Arnone, K. (2023). *Perceptions and Barriers to Adopting Artificial Intelligence in K-12 Education: A Survey of Educators in Fifty States.* Reimagining Education - The Role of E-Learning, Creativity, and Technology in the Post-Pandemic Era., DOI: 10.5772/intechopen.1002741

World Health Organization. (2019). *Para crecer sanos, los niños tienen que pasar menos tiempo sentados y jugar más.*

Yu, H., & Mi, Y. (2023). Application Model for Innovative Sports Practice Teaching in Colleges Using Internet of Things and Artificial Intelligence. *Electronics (Basel), 12*(4), 874. DOI: 10.3390/electronics12040874

Zha, H., Li, W., Wang, W., & Xiao, J. (2025a). The Paradox of AI Empowerment in Primary School Physical Education: Why Technology May Hinder, Not Help, Teaching Efficiency. *Behavioral Sciences (Basel, Switzerland), 15*(2), 240. DOI: 10.3390/bs15020240 PMID: 40001871

Zha, H., Li, W., Wang, W., & Xiao, J. (2025b). The Paradox of AI Empowerment in Primary School Physical Education: Why Technology May Hinder, Not Help, Teaching Efficiency. *Behavioral Sciences 2025, Vol. 15, Page 240, 15*(2), 240. DOI: 10.3390/bs15020240

Zhao, Z., & Xing, X. (2024). Design of an AI-Enhanced Medical Research-Informed Multi-Sensor-Based Online Physical Education Motion Capture System. *Computer-Aided Design and Applications*, 253–267. DOI: 10.14733/cadaps.2024.S24.253-267

Zhou, T., Wu, X., Wang, Y., Wang, Y., & Zhang, S. (2024). Application of artificial intelligence in physical education: A systematic review. *Education and Information Technologies, 29*(7), 8203–8220. DOI: 10.1007/s10639-023-12128-2

Zsarnoczky-Dulhazi, F., Agod, S., Szarka, S., Tuza, K., & Kopper, B. (2024). AI BASED MOTION ANALYSIS SOFTWARE FOR SPORT AND PHYSICAL THERAPY ASSESSMENT. *Revista Brasileira de Medicina Do Esporte, 30*. https://doi.org/DOI: 10.1590/1517-8692202430012022_0020I

# Chapter 4
# Green or Sustainable Advertising:
## A Descriptive Analysis

**María Patricia Soroa de Carlos**
https://orcid.org/0000-0002-0917-7940
*UNIE University, Spain*

**Javier Saiz Briones**
*UNIE University, Spain*

## ABSTRACT

*Growing awareness of environmental issues has driven a change in the attitudes and behaviors of consumers, who are demanding more sustainable products and services. In this context, green or sustainable advertising emerges as a key tool for companies seeking to communicate their commitment to the environment and attract a public increasingly concerned about sustainability. The present research focuses on the descriptive analysis of green or sustainable advertising, exploring its characteristics, strategies, trends and challenges, with the aim of understanding its role in promoting more responsible consumption and building a more sustainable future. Some companies focus on highlighting the environmental benefits of their products, such as reducing emissions, using renewable energy or recycling materials. Other companies choose to communicate their sustainable business practices, such as implementing waste reduction programs, supporting conservation projects or committing to fair trade.*

DOI: 10.4018/979-8-3373-2802-7.ch004

## INTRODUCTION

The growing and increasingly pressing global awareness of environmental challenges has catalyzed a profound transformation in consumer attitudes and behaviors. Driven by the urgency to mitigate climate change, biodiversity loss and rampant pollution, citizens are demanding, with an increasingly louder voice, products and services that not only satisfy their functional and aspirational needs, but also minimize their ecological footprint (Kotler & Keller, 2016). This paradigm shift in consumer expectations has led to the emergence and consolidation of green advertising as a fundamental strategic tool for companies seeking to communicate their commitment to sustainability and capture the attention of an increasingly sensitized public (Peattie, 2010).

The emergence of green advertising, far from being a passing fad, has become a complex and multifaceted phenomenon, intrinsically linked to the evolution of green marketing and Corporate Social Responsibility (CSR). Green marketing, as conceived by Ottman (2011), transcends the mere promotion of environmentally friendly products, encompassing a comprehensive business philosophy that integrates environmental considerations at all stages of the product life cycle, from design and production to distribution and disposal. Within this framework, green advertising acts as the communicating arm, translating sustainability initiatives into persuasive messages aimed at consumers, seeking to influence their purchasing decisions and encourage responsible consumption.

The origins of green advertising can be traced back to the growing environmental concerns that emerged strongly at the end of the 20th century, driven by the publication of seminal reports such as The Limits to Growth (Meadows et al., 1972) and the increasing visibility of problems such as pollution and natural resource degradation (Kilbourne, 2004). Initially, advertising with green overtones was sporadic and often reactive, responding to specific environmental crises or pressure from activist groups (Crane, 2000).

However, the 1980s marked a turning point with growing awareness of the ozone hole and acid rain, leading some companies to highlight environmental attributes of their products as a competitive advantage (Ottman, 2011). This period was characterized by messages that were often simplistic and focused on a single environmental benefit, with a focus more on promoting "less bad" products than inherently sustainable (Peattie, 2010).

The 1990s witnessed greater sophistication, driven by the Rio Earth Summit (1992) and the increasing internalization of sustainability as a relevant factor in the purchasing decisions of an increasingly informed consumer segment (Banerjee, 2020). Companies began to integrate environmental concerns into their marketing strategies, although criticism persisted about the lack of transparency and the proliferation of unsubstantiated claims.

In the 21st century, green advertising has experienced significant expansion, driven by the urgency of climate change and the ubiquity of environmental information through the internet and social media (Dahlén et al., 2010). Consumer pressure for authenticity and transparency has forced companies to adopt more holistic approaches, linking their advertising messages to broader corporate sustainability commitments. Thus, contemporary green advertising seeks to construct a shared value narrative, where environmental benefits are presented in line with benefits for the consumer and society at large (Prieto-Castillo et al., 2021).

The proliferation and sophistication of this trend has intensified markedly in the 21st century, driven by the increased visibility of environmental issues through the media and social networks, as well as by growing regulatory pressure and citizen activism (Kilbourne, 2004). Green advertising has gone from being a niche strategy, adopted mainly by companies with an intrinsic commitment to the environment, to become an increasingly widespread practice in a variety of sectors, from food and energy to fashion and transportation (Crane, 2000).

However, this expansion has not been without controversy and challenges. The line between honest communication of sustainability efforts and greenwashing, the practice of communicating misleading or exaggerated information about environmental credentials, has become a critical point of analysis (Lyon & Montgomery, 2015). The proliferation of "green" messages has generated growing skepticism among consumers, who demand transparency, authenticity and verifiable evidence to back up companies' environmental claims (Dahlén et al., 2010).

Studies explore a wide range of issues, including the effectiveness of different green communication strategies in building brand image and consumer loyalty, the influence of environmental seals and certifications on message credibility, the role of emotions and values in consumer response to green advertising, and the ethical and regulatory challenges associated with greenwashing.

In this context of increasing relevance and complexity, this introduction lays the groundwork for an in-depth exploration of the emergence of green advertising. Through the analysis of contemporary academic literature, we seek to understand the distinctive characteristics of this form of communication, the strategies most commonly employed by companies, the emerging trends in its development and the crucial challenges it faces in its quest to promote responsible consumption and contribute to building a more sustainable future. Understanding these elements is essential for companies aspiring to build a solid reputation in a marketplace increasingly aware of the environmental impact of their decisions and for consumers seeking to make informed purchasing decisions consistent with their sustainability values.

## DEFINITION OF GREEN ADVERTISING AND MAIN STRATEGIES

The growing and pressing global awareness of the climate crisis and environmental degradation has deeply permeated the attitudes and behaviors of consumers, who are increasingly demanding products and services that integrate sustainability into their value proposition (Kotler & Keller, 2016; Prieto-Castillo et al., 2021). In this context, green advertising has emerged as a field of study and a business practice of growing relevance, seeking to communicate the environmental commitment of organizations and connect with a sensitized public (Peattie, 2010; Leonidou et al., 2020).

Defining green advertising requires an understanding that transcends the mere promotion of environmentally friendly products. At its core, green advertising refers to any form of marketing communication that explicitly or implicitly informs, persuades or reminds consumers about the environmental benefits of a product, service, brand or corporate activity (Banerjee, 2020; Dahlén et al., 2010). This definition covers a broad spectrum of messages, from those that highlight specific ecological attributes of a product (such as its recyclability or organic origin) to broader campaigns that seek to build a brand image associated with environmental responsibility and sustainability (Crane, 2000).

Recent academic literature in Spanish and English since 2020 underscores the complexity and evolution of green advertising. Effective green advertising must go beyond simple slogans, integrating authentic and transparent narratives that demonstrate a genuine commitment to sustainable practices across the value chain. Prieto-Castillo et al. (2021) explore how green advertising can act as a catalyst for responsible consumption, as long as it is based on credible information and avoids falling into greenwashing tactics.

In the Anglo-Saxon field, Leonidou et al. (2020) offer a comprehensive perspective, defining green advertising as a subset of green marketing that focuses specifically on communication. These authors stress the importance of congruence between advertising promises and actual company actions, warning about the detrimental effects of greenwashing on consumer trust and brand reputation. Banerjee (2020) stresses that successful green advertising must resonate with the values and environmental concerns of the target audience, using clear language and avoiding ambiguity that can generate skepticism.

In addition, green advertising is increasingly influenced by new technologies and digital channels. Social media, content marketing and interactive advertising offer new opportunities to communicate sustainability messages in a more direct and participatory way (Dahlen et al., 2010). However, these new formats also present challenges in terms of transparency and the need to combat environmental misinformation.

Contemporary green advertising is defined as a multifaceted communication strategy that seeks to inform, persuade and remind consumers about the environmental benefits associated with a company's offerings and practices. Its effectiveness and credibility depend crucially on authenticity, transparency and consistency with a real commitment to sustainability, avoiding the misleading practices of greenwashing and genuinely resonating with the values of a consumer increasingly aware of the future of the planet. Ongoing research continues to explore the best practices and ethical challenges of this evolving field.

The main characteristics of green advertising are, on the one hand, that it has a) a focus on environmental benefits - explicit or implicit promotion of the ecological attributes of a product, service, brand or business activity. This may include emissions reduction, renewable energy use, energy efficiency, sustainable sourcing of materials, biodegradability, recyclability, waste reduction, natural resource conservation or biodiversity support (Banerjee, 2020; Leonidou et al., 2020)-. In addition, b) attempt to persuade towards sustainable behaviors-it seeks to influence consumer attitudes and behaviors towards more sustainable choices. This may involve encouraging the purchase of green products, promoting responsible use practices, or even advocating changes in consumption habits (Peattie, 2010; Prieto-Castillo et al., 2021). c) Emphasis on credibility and transparency: in the face of growing skepticism about greenwashing, credibility and transparency have become essential features. Effective green advertising relies on verifiable information, hard data, independent certifications, and clear and honest communication about the company's environmental efforts (Dahlén et al., 2010; Font et al., 2020). d) Connection with consumer values and emotions: successful green advertising often seeks to resonate with the values and emotions of environmentally concerned consumers. This may involve appealing to feelings of responsibility, connection to nature, concern for the future, or a desire to contribute to a better world (Kotler & Keller, 2016). e) Integration with Green Marketing Strategy and CSR: effective green advertising is aligned with the company's overall green marketing strategy and Corporate Social Responsibility (CSR) commitments. It should reflect actual sustainable actions and practices throughout the organization (Banerjee, 2020; Leonidou et al., 2020). And f) Adaptation to various channels and formats: green advertising manifests itself in a wide range of channels and formats, from traditional print and broadcast ads to digital campaigns, content marketing, social media advertising, influencer partnerships and interactive experiences (Dahlén et al., 2010).

Research on green advertising has employed various methodological perspectives to address its complexity and multiple dimensions. Based on the current academic literature, we can identify at least five main methodological perspectives:

*Table 1. Appropriate research methods*

| Method | Description | Objectives | Interest |
|---|---|---|---|
| **Content analysis** | This perspective focuses on the systematic and objective examination of the overt content of green advertisements. Researchers analyze textual and visual messages, the use of symbols and images related to nature, mentions of specific environmental attributes, persuasive appeals used, and the presence of certifications or eco-labels. | Identify the predominant communication strategies, trends in the representation of sustainability, the prevalence of certain environmental issues and the possible presence of greenwashing elements. | It provides a macro view of the communication strategies and trends employed by companies. Identifies what is being said and how sustainability is being presented. Focuses on the message level |
| **Consumer reception and perception studies** | This perspective focuses on understanding how consumers consumers process, interpret and respond to green advertising messages. Quantitative (surveys, experiments) and qualitative (focus groups, in-depth interviews) methods are used to assess consumers' attitudes, beliefs, purchase intentions and level of skepticism towards green advertising. | Evaluate the effectiveness of different types of green messages, identify factors influencing the perceived credibility of green advertising, understand the impact of green advertising on brand image and consumer loyalty, and analyze the emotional and cognitive responses of consumers to green advertising. perceived credibility of green advertising, understand the impact of green advertising on brand image and consumer loyalty, and analyze the emotional and cognitive responses of consumers. | They provide a micro-understanding of how consumers interpret and react to those messages. They allow you to assess whether the communication is being effective and how it influences their attitudes and behaviors. Focus on the level of the individual (the consumer). |
| **Case studies** | This perspective involves the in-depth analysis of specific green advertising campaigns of one or more companies. Qualitative and sometimes quantitative data are collected on campaign strategy, implementation, public response and results obtained. | Gain a detailed understanding of the factors that contribute to the success or failure of green advertising campaigns, identify best practices, and draw lessons learned for future initiatives | They allow in-depth analysis of specific campaigns, revealing the complexities and contextual factors that influence their success or failure. Are at an intermediate level, analyzing specific campaigns in context.. |

continued on following page

*Table 1. Continued*

| Method | Description | Objectives | Interest |
|---|---|---|---|
| **Experimental approaches** | This perspective uses experimental designs to manipulate specific variables of green advertisements (e.g., type of environmental storyline, presence of certifications, message tone) and measure their impact on consumer responses (attitudes, purchase intentions, perceived credibility). | Establish causal relationships between the characteristics of green advertising and consumer responses, identify the most persuasive elements and understand the psychological mechanisms underlying the effectiveness of green advertising. | They offer the possibility of establishing causal relationships between specific elements of green advertising and consumer responses, providing empirical evidence on persuasion. they focus on the level of the individual, but under controlled conditions. |
| **Longitudinal and trend analysis** | This perspective examines the evolution of green green advertising over time, analyzing how the strategies, themes, and sophistication of green messages have changed. It may also involve analyzing trends in consumer response to green advertising over different time periods. | Identify emerging trends in sustainability communication, understand how green advertising is adapting to changes in environmental awareness and consumer expectations, and assess the long-term effectiveness of different strategies. and consumer expectations, and evaluate the long-term effectiveness of different strategies. | It allows to understand the evolution of the phenomenon over time, identifying changes and adaptations in strategies and reception. It is situated at a macro level, observing the evolution of the phenomenon at a social or industrial level. |

Source: prepared by the author based on different sources.

The table presented shows five fundamental methodological perspectives for approaching research in the field of green advertising, detailing for each one its central method, the description of its approach, the primary objectives it pursues and the interest or distinctive contribution it brings to the field of study. These perspectives, far from being mutually exclusive, offer a comprehensive framework for analyzing the complexity of sustainability communication from different angles and levels of analysis.

The first perspective, content analysis, stands as a systematic and objective method for examining the textual and visual corpus of green advertisements. Its description emphasizes the exploration of overt messages, including the use of symbols, images, appeals and labels related to environmental attributes. The central objective lies in identifying the predominant communicative strategies, trends in the representation of sustainability, the prevalence of certain themes and the possible presence of greenwashing. The interest of this methodology lies in providing a panoramic view of what is being communicated and how sustainability is being presented at the message level, allowing the identification of general patterns and trends in the advertising industry.

The second perspective, consumer reception and perception studies, focuses on the opposite pole of the communicative process: the audience. Its description emphasizes the investigation of how consumers interpret and respond to green advertising messages, employing both quantitative (surveys, experiments) and qualitative (focus groups, in-depth interviews) methods. The primary objectives are to assess the effectiveness of different types of messages, identify factors influencing perceived credibility, understand the impact on brand image and loyalty, and analyze consumers' emotional and cognitive responses. The interest of this perspective lies in providing a micro understanding of how individuals process green advertising, allowing to discern its impact on attitudes, beliefs and purchase intentions at the individual level.

The third perspective, case studies, takes an in-depth approach to analyze specific campaigns or the green advertising strategy of individual companies. Its description emphasizes the collection of qualitative and sometimes quantitative data, focusing on the planning, implementation and results of specific campaigns. The main objective is to gain a detailed understanding of the factors contributing to the success or failure of these initiatives, identify best practices and extract lessons learned for future implementations. The interest of this methodology lies in providing a contextualized and richly detailed analysis of green advertising in practice, revealing the complexities and nuances that influence its effectiveness at the organizational and campaign level.

The fourth perspective, experimental approaches, is characterized by the controlled manipulation of specific variables of green advertisements (e.g., type of environmental claim, presence of certifications, message tone) to measure their impact on consumer responses (e.g., attitudes, purchase intentions, perceived credibility). Its description emphasizes the design of causal studies to establish direct relationships between message elements and audience reactions. The fundamental objective is to establish clear causal relationships and to understand the psychological mechanisms underlying the effectiveness of green advertising. The interest of this methodology lies in providing rigorous empirical evidence on the factors that influence consumer responses under controlled conditions, allowing to isolate variables and determine cause and effect relationships at the individual level.

Finally, the fifth perspective, longitudinal and trend analysis, takes a macro and temporal view to examine the evolution of green advertising over time. Its description emphasizes the analysis of how the strategies, themes and sophistication of green messages have changed, as well as the possible influence of social and economic trends. The main objectives are to identify emerging trends in sustainability communication, to understand how green advertising adapts to changes in regulation and consumer expectations, and to assess its long-term effectiveness. The interest of this methodology lies in providing a dynamic and evolutionary understanding of the phenomenon, observing changes and adaptations at the societal or industry level.

## PERCEPTION AND RELEVANCE OF GREEN ADVERTISING

The perception and effects of green advertising on consumers is a complex and multifaceted area of research within marketing and communication studies. The effectiveness of environmental messages is not uniform and is modulated by a variety of individual, contextual and message factors (Leonidou et al., 2020). The perceived credibility of green advertising emerges as a critical factor. Consumers, increasingly aware of the potential for greenwashing, evaluate the authenticity of environmental claims based on the transparency of information, the presence of independent certifications, and the consistency between the advertising message and actual corporate actions (Font et al., 2020).

A significant factor influencing perception is the consumer's level of environmental involvement. Those with a high level of environmental concern tend to be more receptive to green advertising, but also more skeptical of vague or unsubstantiated claims, demanding detailed and verifiable information. On the other hand, consumers with less involvement may be influenced by peripheral cues such as the use of natural images or colors associated with ecology, although their information processing is less profound (Dahlén et al., 2010).

In terms of effects, green advertising can impact various dimensions. At the cognitive level, it can increase consumer knowledge about products' environmental attributes and companies' sustainable practices. At the affective level, it can generate positive emotions such as admiration or connection with the brand, strengthening corporate image and consumer loyalty (Prieto-Castillo et al., 2021). At the conative level, the ultimate goal is to influence purchase intention and consumer behavior, promoting the choice of more sustainable products and services.

However, positive effects are not automatic. Poorly executed green advertising or perceived as greenwashing can generate skepticism, cynicism, and a decrease in trust in the brand and even in environmental communication as a whole (Lyon & Montgomery, 2015). The incongruity between advertising promises and the reality of business practices can have lasting negative consequences for corporate reputation.

The perception and effects of green advertising are determined by the credibility of the message, the level of consumer environmental engagement, and the consistency between the communication and the company's actions. Authentic and transparent green advertising has the potential to positively influence consumer attitudes, knowledge, and behavior, contributing to more responsible consumption. However, greenwashing represents a significant risk that can erode trust and undermine the objectives of sustainability communication.

## THE DANGER OF GREENWASHING: BAD PRACTICE IN GREEN ADVERTISING

Greenwashing is emerging as a significant risk and a pernicious consequence of poor green advertising practices by brands. Defined as the misleading or exaggerated communication of the environmental benefits of a product, service, or corporate practice (Lyon & Montgomery, 2015), greenwashing undermines the credibility of both the offending brand and sustainability communication as a whole, generating skepticism and cynicism among consumers (Dahlén et al., 2010).

The temptation of greenwashing arises from the growing demand for sustainable products and services from consumers increasingly aware of the environmental impact of their choices (Kotler & Keller, 2016). Brands, seeking to capitalize on this trend, may resort to superficial or downright deceptive communication strategies to project an image of ecological responsibility without implementing substantial changes in their operations or the composition of their products (Banerjee, 2020).

The manifestations of greenwashing are diverse, including the use of vague and undefined terms ("natural", "ecological"), the exaggeration of minor environmental benefits while ignoring significant impacts, the presentation of selective or unverified information, the use of false or misleading seals or certifications, and the creation of "green" images evocative of nature without any real backing (TerraChoice Environmental Marketing, 2007).

The danger of greenwashing for brands is multifaceted. In the short term, it can generate an increase in sales by attracting consumers with sustainable purchasing intentions. However, in the medium and long term, the discovery of deceptive practices can have serious consequences for corporate reputation. The loss of consumer trust is significant and difficult to repair, and can translate into boycotts, negative publicity, and an erosion of brand loyalty (Font et al., 2020).

Furthermore, greenwashing can lead to legal and regulatory consequences. Consumer protection authorities in various jurisdictions are intensifying their scrutiny of companies' environmental claims, imposing fines and requiring redress in cases of deceptive advertising (Federal Trade Commission, 2012). Reputational damage and financial penalties can have a significant impact on the viability and value of the brand.

Finally, greenwashing has a detrimental effect on progress toward true sustainability. By confusing consumers and making it difficult to identify genuinely sustainable brands and products, it hinders the development of a market based on transparency and environmental responsibility (Lyon & Montgomery, 2015). The distrust generated by the poor practices of some brands can extend to the entire green marketing sector, hindering effective communication by companies with a genuine commitment to sustainability.

Greenwashing represents a significant danger for brands that engage in poor green advertising practices. While it may offer short-term benefits, the long-term consequences in terms of lost consumer trust, reputational damage, legal penalties, and contributing to widespread distrust in sustainability communication far outweigh any initial gains. Authenticity, transparency, and verification of environmental claims are crucial to building effective and ethical green advertising, which benefits both brands and the movement toward a more sustainable future.

## EFFECTS OF GREEN ADVERTISING ON THE CONSUMER

The effects of green advertising on consumers constitute a significant area of research within marketing and communication, exploring how environmental messages influence their attitudes, beliefs, purchase intentions and actual behaviors (Leonidou et al., 2020). The academic literature suggests that these effects are complex and mediated by a variety of consumer, message, and context-related factors (Dahlén et al., 2010).

From a cognitive standpoint, green advertising can increase consumer awareness and knowledge of environmental issues, green product attributes, and sustainable business practices (Prieto-Castillo et al., 2021). Clear messages and specific information about environmental benefits (e.g., emissions reduction, energy efficiency, use of recycled materials) can educate consumers and improve their understanding of available sustainable options (Font et al., 2020). However, information overload or the presentation of complex data can lead to confusion or even rejection.

In the affective sphere, green advertising has the potential to evoke positive emotions such as a connection with nature, concern for the future of the planet or the feeling of contributing to a noble cause. These emotional responses can strengthen the consumer's connection to the brand and generate favorable attitudes toward products and companies that are perceived as environmentally responsible. Narratives that are authentic and appeal to personal sustainability values tend to be more effective in this regard.

At a conative level, the primary objective of green advertising is to influence purchase intent and, ultimately, consumption behavior. Credible and relevant green advertising can motivate consumers to choose more sustainable products and services, even if they involve a slightly higher cost (Kotler & Keller, 2016). However, intention does not always translate into action, and factors such as availability, price, convenience, and social norms also play a crucial role in the final purchase decision.

Nevertheless, the effects of green advertising are not inherently positive. Perceived inconsistency or lack of authenticity, especially when consumers suspect greenwashing, can lead to skepticism, cynicism, and decreased trust in the brand and in environmental communication in general (Lyon & Montgomery, 2015). Misleading or exaggerated advertising can have a lasting negative impact on a company's reputation and erode the credibility of the entire green marketing industry.

In addition, the effects of green advertising can vary significantly among different consumer segments. Those with high environmental involvement and prior knowledge about sustainability tend to be more receptive and critical of messages, while consumers with less involvement may be influenced by more superficial cues (Dahlén et al., 2010).

Green advertising has the potential to generate positive consumer effects at the cognitive, affective and conative levels, fostering awareness, favorable attitudes and purchase intentions towards sustainable options. However, credibility, authenticity and relevance of the message are crucial to avoid skepticism and cynicism associated with greenwashing. Understanding the diverse responses of different consumer segments is critical to designing effective green advertising campaigns that contribute to more responsible consumption and building a more sustainable future.

More specifically, at a cognitive level, one of the main advantages is the potential to increase consumer awareness and knowledge about specific environmental issues, the ecological attributes of products and the actions they can take to mitigate their impact. Informative and transparent campaigns can educate about the importance of recycling, energy efficiency or responsible consumption, empowering the consumer with relevant information to make more informed decisions. However, a significant disadvantage lies in information overload or the presentation of complex and contradictory data, which can lead to confusion, information fatigue and even skepticism towards environmental claims. In addition, the proliferation of superficial or misleading "greenwashing" messages can erode the credibility of all sustainability communication, making it difficult for consumers to distinguish between genuine initiatives and opportunistic marketing strategies.

On the affective level, well-executed green advertising has the advantage of evoking positive emotions such as a connection with nature, concern for the future of the planet and a feeling of contributing to a noble cause. These emotional responses can strengthen the consumer's connection to the brand, generate favorable attitudes and foster a sense of shared identity around sustainability values. However, a disadvantage emerges when green advertising is perceived as manipulative, exploiting consumer fear or guilt without offering real solutions, or when the company's actions do not match its messages, generating frustration, cynicism and even rejection of the brand. An excessively alarmist or moralizing tone can also alienate certain segments of the public.

From a conative point of view, the main advantage lies in the ability of green advertising to influence purchase intent and, ultimately, consumer behavior. Persuasive messages that highlight the functional, economic or social benefits of sustainable choices, in addition to the environmental ones, can motivate the choice of greener products and services. However, a major drawback is that intent does not always translate into action. Factors such as price, availability, convenience, and social norms often trump environmental considerations in the final purchase decision. In addition, green advertising that fails to overcome the perception of a sacrifice in quality or performance may not achieve the desired impact on consumer behavior. Finally, greenwashing can generate widespread distrust that hinders the adoption of sustainable consumption behaviors even when genuine alternatives are offered.

## ETHICS AND GREEN ADVERTISING

The intersection between ethics and green advertising is fertile ground for academic study, revealing paradoxes and challenges that go beyond the mere communication of environmental attributes. One of the most notable curiosities is the transparency paradox, where the growing consumer demand for authenticity often meets the complexity of global supply chains and the difficulty of verifying all environmental claims. Although companies strive to communicate their sustainability initiatives, the asymmetry of information and the very opacity of some industrial processes can lead to skepticism, even in the face of genuine efforts (Lyon & Montgomery, 2015).

Green advertising has become a crucial strategic tool for brands seeking to connect with increasingly sustainability-conscious consumers (Peattie, 2010). However, its increasing prevalence has brought with it significant ethical and regulatory challenges, primarily focused on the risk of greenwashing - the practice of communicating misleading or exaggerated information about environmental credentials (Lyon & Montgomery, 2015).

Another fascinating aspect lies in the protean nature of greenwashing. Far from being a monolithic practice, greenwashing manifests itself in a variety of forms, from selective omission of information to exaggeration of minor benefits, to the creation of misleading labels or distraction from the true environmental impact (Lyon & Montgomery, 2015). This multiplicity makes its detection a constant challenge for consumers and regulators. The curiosity here lies in how this practice adapts and evolves in response to public scrutiny and changing regulations, driving a "cat-and-mouse game" between companies and sustainability watchdogs.

It is also intriguing how the burden of ethical proof falls asymmetrically between the company and the consumer. Often, the consumer is expected to invest time and effort in verifying the authenticity of green claims, despite the fact that companies are the ones in possession of complete information about their processes and products. This situation underscores the need for stricter regulatory frameworks and clear disclosure standards that shift the primary responsibility for truthfulness to the one issuing the message (Peattie, 2010). Curiosity extends to the effectiveness of third-party certifications: although they seek to generate confidence, the proliferation of seals and labels, sometimes poorly regulated or known, can paradoxically increase consumer confusion.

From an ethical perspective, green advertising has a responsibility to be truthful, transparent and grounded in sound scientific evidence (Gbadamosi, 2019). Brands should avoid using vague or ambiguous terms such as "eco-friendly" or "natural" without clear and verifiable justification (Alvarez-Surera, 2023). Omitting relevant information, exaggerating marginal benefits, and focusing on a single "green" attribute while ignoring significant negative impacts are ethically questionable practices that erode consumer trust. A fundamental ethical principle calls for alignment between advertising messages and a company's actual environmental performance, prioritizing genuine improvement over mere "green cosmetics."

At the regulatory level, several initiatives at the national and international levels seek to combat greenwashing and ensure more honest environmental communication. The European Union, for example, has implemented directives to prevent companies from marketing their products and services as environmentally sustainable without verifiable evidence (European Commission, 2024). These regulations often prohibit the use of generic environmental claims without a demonstration of excellent environmental performance, and require claims to be validated by an independent body. In the United States, the Federal Trade Commission (FTC) has published Green Guides, which, although they do not have the force of law, provide administrative interpretations of the laws against misleading advertising and offer guidance to advertisers on how to avoid making misleading environmental claims (Green Guides, n.d.). These guides address issues such as the use of certifications, recycled content claims, biodegradability and carbon offsetting.

Compliance with these regulations and the adoption of ethical principles in green advertising are not only crucial to avoid legal sanctions and reputational damage, but also to build consumer trust and foster a market for genuinely sustainable products and services. Transparency, accountability and the provision of substantiated evidence are essential elements for ethical and effective green advertising that contributes to more responsible consumption and environmental protection. Ultimately, the credibility of green advertising depends on the consistency between what brands communicate and what they actually do in terms of sustainability.

From an affective perspective, it is curious to note how green advertising, by appealing to intrinsic values such as responsibility and altruism, can generate intense emotional responses that, if perceived as manipulated or inauthentic, can turn against the brand. "Green guilt" or "climate anxiety" are emotions that advertising might inadvertently exploit, but which can also lead to fatigue or cynicism if messages lack a basis of credibility (Leonidou et al., 2020). This highlights the fine ethical line between inspiring positive action and generating distrust through irresponsible communication.

Finally, the "least bad" dilemma is a recurring curiosity. In industries with an inherently high environmental impact (e.g., fast fashion, packaged beverages), green advertising often focuses on incremental improvements ("less plastic," "more sustainable cotton") rather than addressing systemic problems or promoting a decrease in consumption. The ethics here lie in whether these communications, while technically true, divert attention from the need for deeper transformations in the business model (Ottman, 2011). This is a constant point of debate in the literature, which questions the extent to which current green advertising really contributes to a sustainable transition or whether, in some cases, it perpetuates consumption patterns through environmental "image washing".

## RESEARCH OBJECTIVES

Growing global environmental awareness has prompted companies to integrate sustainability into their marketing strategies, leading to the proliferation of green advertising. This research sets out to take an in-depth look at this multifaceted phenomenon. This research seeks to provide a comprehensive view of the current state and complexities of green advertising. Therefore, the objectives of this paper are as follows:

- Describe the characteristics and trends of green or sustainable advertising.
- To analyze the main green advertising campaigns since 2020.
- Identify and categorize the green advertising strategies most commonly employed by companies.
- Evaluate the effectiveness of green advertising in building brand image and consumer loyalty
- Analyze the ethical and regulatory challenges associated with greenwashing in green advertising.

## METHODOLOGY

This research adopts a descriptive methodology with the aim of analyzing the characteristics, strategies and predominant trends in contemporary green advertising. Given the exploratory nature and the purpose of offering a detailed view of the phenomenon, a methodological approach has been chosen that prioritizes the collection and systematic analysis of advertising content, complemented by the review of relevant academic literature.

The selection of the sample of advertisements will be carried out through a non-probabilistic purposive sampling, seeking to cover a diversity of industrial sectors (food, automotive, fashion, energy, personal care, among others) and advertising formats (television, print, online advertising, social networks). Priority will be given to advertising campaigns broadcast or published during the period 2020-2025, in order to reflect current trends in sustainability communication. The selection will be based on criteria of visibility, reach and perceived relevance in the public discourse on sustainability, identified through media analysis and online monitoring tools.

Data collection will be carried out through documentary analysis of the selected advertising content. Key variables related to the characteristics of the advertisement will be recorded and coded, such as the type of product or service promoted, the dissemination medium, the presence and prominence of specific environmental claims, the use of visual and textual elements associated with sustainability (images of nature, green colors, terms such as "ecological", "natural", "sustainable"), the mention of environmental certifications or labels, and the predominant persuasive strategy (emphasis on product attributes, production processes, corporate social responsibility).

Data analysis will be performed using a qualitative content analysis approach. Recurrent patterns, predominant communicative strategies and possible cases of greenwashing will be identified through the evaluation of the clarity, specificity and verifiability of environmental claims, following the criteria proposed by theoretical frameworks on greenwashing (Lyon & Montgomery, 2015) and regulatory guidelines (FTC Green Guides, European Union). We will seek to categorize the identified green advertising strategies (e.g., emphasis on minor attributes, use of vague language, corporate image whitewashing) and analyze their frequency and distribution by sector and advertising format.

To complement the analysis of advertising content, an exhaustive literature review of the academic literature published in journals indexed in JCR and Scopus during the same period (2020-2025) will be conducted. This review will focus on studies addressing the definition, characteristics, strategies, consumer perception, ethical and regulatory aspects of green advertising. The objective of this review is to contextualize the findings of the content analysis, identify relevant theoretical frameworks, and contrast observed advertising practices with academic recommendations and analyses.

The interpretation of the results will be done by triangulating the data obtained from the content analysis and the literature review. We will seek to identify the dominant trends in green advertising, the strategies most used by companies, the coherence between advertising practices and academic discourse on sustainability communication, and the possible ethical and regulatory challenges manifested in the analyzed sample. The presentation of the results will be descriptive, using concrete examples drawn from the content analysis to illustrate the characteristics and strategies identified.

## RESULTS

In this section, we will analyze six multinationals that for more than five years have carried out and continue to use green marketing in their campaigns. To do so, we have relied not only on academic works, but also on their own websites where we have been able to consult both their advertising campaigns in different online and offline media.

In order to study it in depth, in the following table we present the main findings:

*Table 2. Analysis of the advertising campaigns of six multinational brands*

| Brand | Industry | Campaign | Green advertising practices | Use of green advertising words | Green marketing policies | Tone of communication |
|---|---|---|---|---|---|---|
| Patagonia | Fashion/ textile | "1% for the Planet" (2002) | It stands out for its commitment to sustainability and social responsibility, implementing programs to reduce its carbon footprint, promoting the reuse and recycling of products, and donating 1% of its sales to environmental organizations. | "Long-lasting", "Conservation", "Donation to environmental causes" | Product Life Extension, Partnerships with NGOs and Environmental Organizations | Patagonia tends to be more direct and activist in its tone. |
| Tesla | Automotive | Does not do traditional advertising | It has revolutionized the automotive industry by making electric vehicles an attractive option, reducing greenhouse gas emissions and driving clean energy technologies. | "Energy efficiency", "Zero emissions", "Renewable energy" | "Green" Products and Services | Tesla focuses on innovation and mission |
| Unilever | Hygiene/cleaning | "Sustainable Living Plan" (2010) | With its Sustainable Life Plan, it connects its line of business with respect for the environment, recovering used packaging for recycling, promoting the circular economy, and making | "Organic", "Sustainably Sourced", "Biodegradable/ Compostable", "Reusable", "Circular Economy" | "Green" Products and Services, Sustainability Narrative | Unilever tends to have a more corporate tone and mass reach. |
| Coca-Cola | Food / drinks | "World Without Waste" (2018) | It promotes recycling and reusing bottles, as seen in its "Share a Coke" campaign. | "Recycled/ Recyclable", "Reusable", "Recycling Programs" | Promotion of Recycling and Reuse | For the general public with a corporate language |
| IKEA | Furniture/ decoration | "People & Planet Positive" (2012) | Recognized for its efforts to contribute to environmental conservation, it manages waste, renews energy with solar panels and wind farms, and has planted millions of trees. | "Recycled/ Recyclable", "Sustainably Sourced", "Energy Efficiency", "Circular Economy", "Renewable Energy" | Sustainable Design, Sustainability Narrative | IKEA takes a more practical, home-oriented approach. |
| H&M | Fashion/ textile | "Conscious" (2010) | It promotes sustainable practices and rewards users who take sustainable actions, such as recycling clothing. | "Recycled/ Recyclable", "Sustainably Sourced", "Circular Economy", "Recycling Programs" | Green Products and Services, Promotion of Recycling and Reuse | H&M often uses a more accessible and fashionable tone |

Source: prepared by the author based on different sources.

The comparative table presents a descriptive analysis of the green marketing strategies employed by six leading companies in various sectors: Patagonia (fashion/textiles), Tesla (automotive), Unilever (hygiene/cleaning), Coca-Cola (food/beverage), IKEA (furniture/design) and H&M (fashion/textiles). The analysis is articulated around five key dimensions: the core green marketing campaign, the underlying green advertising practices, the use of specific keywords, the green marketing policies implemented and the tone of the communication adopted.

Regarding the core campaign, a diversity of approaches is observed. Patagonia focuses its communication on its founding commitment to donate 1% of its sales to environmental causes ("1% for the Planet"), establishing a direct connection between the consumption of its products and support for conservation. Unilever, for its part, implemented a long-term strategic framework ("Sustainable Living Plan") that integrates sustainability into various facets of its business and communications. Coca-Cola focuses its green advertising on packaging waste management ("World Without Waste"), addressing a significant environmental challenge in its sector. IKEA articulates its strategy around a holistic concept ("People & Planet Positive") that encompasses everything from product design to store operations. H&M promotes a specific product line ("Conscious") with the promise of using more sustainable materials. Tesla differentiates itself by not using traditional advertising, basing its marketing on technological innovation and its mission to accelerate the transition to sustainable energy.

The underlying green advertising practices reveal a convergence in the communication of concrete actions. Patagonia highlights the durability of its products, the promotion of repair, and direct donation. Tesla emphasizes energy efficiency, the absence of direct emissions, and the use of renewable energy. Unilever communicates circular economy initiatives and the use of sustainably sourced ingredients. Coca-Cola focuses on the recycling and reuse of its packaging. IKEA promotes sustainable design, energy efficiency, and the use of renewable energy. H&M communicates the use of recycled materials and clothing recycling programs.

The use of specific keywords illustrates each company's communication priorities. Patagonia uses terms such as "durable," "conservation," and "donation," reinforcing its brand image as committed to long-term environmental protection. Tesla relies on technical terms such as "energy efficiency," "zero emissions," and "renewable energy," appealing to the innovation and functionality of its vehicles. Unilever uses a broad vocabulary that includes "organic," "biodegradable," "recyclable," and "circular economy," reflecting the diversity of its products and commitments. Coca-Cola focuses on "recycled" and "recycling programs," aligning with its core campaign. IKEA uses terms like "upcycled," "sustainably sourced," "energy efficiency," and "renewable energy," communicating its holistic approach. H&M uses similar language to IKEA and Coca-Cola, emphasizing "recycled" and "circular economy," though with a particular focus on affordable fashion.

The green marketing policies adopted vary in scope and emphasis. Patagonia integrates sustainability into the product's lifecycle and collaborates with environmental organizations. Tesla focuses its marketing on the inherent "green" attributes of its products. Unilever adopts a broad-based corporate sustainability narrative. Coca-Cola focuses on promoting recycling and reuse. IKEA articulates a home- and design-oriented sustainability narrative. H&M promotes the recyclability and accessibility of its "green" products.

Finally, the tone of communication reflects each company's brand identity and target audience. Patagonia adopts a direct, activist tone. Tesla focuses on innovation and mission. Unilever maintains a corporate tone with a mass reach. Coca-Cola uses corporate language for a broad audience. IKEA adopts a more practical, home-oriented tone. H&M uses an accessible, trendy tone.

It should be noted that we have found that, unfortunately, four of the six brands analyzed are also accused of greenwashing. Although the sources are not academic, but journalistic in nature, it has resonated with public opinion. For example, H&M has been subject to criticism and accusations of greenwashing regarding its "Conscious" line (Business & Human Rights Resource Centre, 2022). The allegations focus on the fact that, while they promote the use of recycled and organic materials, the proportion of these materials in their overall production remains relatively small. Transparency regarding the origin and actual impact of these materials has also been questioned, suggesting that the "Conscious" label could give an exaggerated impression of its overall sustainability.

Unilever has also faced accusations of greenwashing in relation to its broad sustainability initiatives and its "Sustainable Living Plan" (now "Compass"), related to specific products or marketing campaigns. Some criticism has focused on the lack of ambition for certain goals or the promotion of products that, while having some "green" attributes, still have a significant environmental impact in other areas. There has also been controversy over the advertising of certain products that contain questionable ingredients from an environmental or health perspective, despite being marketed as having a focus on wellness or sustainability (Ethical Consumer, n.d.).

Even Coca-Cola has been accused of malpractice, as its "World Without Waste" campaign focuses on a real problem (packaging waste). The company has been criticized for not aggressively addressing the reduction of virgin plastic production. Accusations of greenwashing often arise when its recycling efforts are contrasted with the vast amount of plastic it continues to produce and place on the market. There have also been controversies over sponsorships of events or initiatives that are presented as sustainable, but do not address the core impacts of their business (BEUC, n.d.).

Although IKEA is generally perceived as a company with a relatively strong commitment to sustainability, it has also faced scrutiny. Some criticism has focused on the sustainability of certain materials or the durability of some of its products, contrasting this with its image as a company that promotes a more conscious lifestyle. There have also been questions about the true scale of its renewable energy initiatives in relation to its global footprint (China Water Risk, 2020).

The only brand that has escaped greenwashing is Patagonia. It is generally considered one of the most authentic companies in its environmental commitment and has faced fewer significant accusations of greenwashing. However, even Patagonia has occasionally faced scrutiny over its entire supply chain and the impact of its clothing production, although its transparency and activism generally generate greater trust. In the area of sustainability, criticism of Tesla is sometimes directed at the carbon footprint of battery production or the origin of the materials used, rather than the "green" marketing of its electric vehicles themselves, whose fundamental environmental benefits are generally recognized.

Patagonia is known for a diverse color palette in its apparel, strongly rooted in mountaineering and nature (including olive greens, forest, etc.), its green marketing goes beyond color. However, in its communications, logos for environmental initiatives (such as "1% for the Planet") and in the design of its brand identity, green is a recurring color that reinforces its message of connection to nature and conservation. It would not be unusual to find specific Patagonia products or packaging with an emphasis on green if they relate to specific "eco" collections or materials.

Tesla focuses primarily on technology and energy efficiency, and its primary branding (logo, vehicles) tends to be more neutral (black, white, silver) or metallic, reflecting its image of innovation and modernity. However, in advertising materials or infographics explaining the environmental benefits of their vehicles (zero emissions, renewable energy), they are very likely to use graphics and schemes with green tones to symbolize sustainability and clean energy. It is not common for your cars or the main packaging of your products (such as the vehicles themselves) to be green by default, but the color would be used to communicate the green impact.

Given Unilever's vast range of hygiene and cleaning products and its multiple brands, the use of the color green is extremely common in those products or lines that want to convey a message of "natural", "eco-friendly" or "clean without chemicals". For example, brands such as Seventh Generation (which is part of Unilever) make extensive use of green in their packaging and branding. It is almost certain that, within its "sustainable" or "eco" lines, Unilever employs the color green prominently in packaging, product-specific logos or campaigns highlighting natural ingredients or sustainable processes.

Historically associated with red, Coca-Cola has made a strategic transition to incorporate green into its sustainability-related marketing. The brand has used and continues to use green bottles and labels for its "Coca-Cola Life" or "Coca-Cola Zero Sugar Green" variants (which have had different iterations), where green communicates a message of "more natural" or "less sugar" ingredients that aligns with a perception of "healthier" or "greener." In addition, in its "World Without Waste" campaign, it is common to see green graphic elements in its advertising materials to symbolize recycling and the circular economy.

IKEA is known for its blue and yellow branding, but in the context of its sustainability products and solutions for the home, green is a widely used color. In its catalogs, stores and campaigns related to energy-efficient products, LED lighting, sustainable cotton textiles, or recycling and waste reduction solutions, the color green features prominently in graphics, product-specific packaging and signage. Its "People & Planet Positive" strategy is also often communicated with green visual elements.

For its "Conscious" line and other sustainability initiatives, H&M actively employs the color green in its stores, garment labels, marketing campaigns and promotional materials. Green is used to visually identify collections that are supposed to have a lower environmental impact, even if the garments themselves are not green. Conscious" line labels often incorporate the color green to signal their "sustainable" nature.

Most of these companies (especially Unilever, Coca-Cola, IKEA and H&M) have directly or indirectly used the color green in their products, packaging or marketing materials for their green advertising. Patagonia incorporates it as part of its brand identity and in its communications, and Tesla uses it in the graphic representation of the environmental benefits of its technology. The color green is a universally recognized visual tool for communicating sustainability and is a common practice in green marketing.

## FUTURE RESEARCH DIRECTIONS

The descriptive research conducted on green advertising reveals a dynamic and constantly evolving field, marked by the growing environmental awareness of the consumer and the strategic response of companies. From the preliminary findings, several lines of future research emerge that could deepen our understanding of this phenomenon and address the complexities inherent in sustainability communication.

One promising line of research focuses on the longitudinal analysis of green advertising strategies. The current study provides a snapshot of the contemporary landscape; however, a diachronic tracking of campaigns and the language employed by companies would identify evolving trends, the adoption of new narratives, and the response of brands to changes in regulation and consumer expectations. It could be investigated whether companies are adopting more sophisticated and transparent approaches over time, or whether patterns of greenwashing persist, adapted to new media contexts.

Second, it is crucial to delve deeper into the effectiveness of different green advertising strategies in building consumer trust and brand loyalty. The descriptive study identifies various tactics (emphasis on certifications, supply chain transparency, promotion of the circular economy); however, empirical research is required to determine which of these strategies are most effective in generating a genuine connection with consumers and encouraging long-term sustainable purchasing behavior. Quantitative and qualitative methodologies could be employed to analyze consumer perception of different types of messages and the influence of perceived credibility on purchase intent and loyalty.

A third line of research focuses on the influence of new media and digital platforms on green advertising. The rise of social media, influencer marketing, and interactive advertising presents new opportunities and challenges for sustainability communication. Future studies could explore how companies use these channels to interact with consumers on environmental issues, the effectiveness of different content formats (short videos, direct, user-generated content) and the impact of transparency and authenticity in these digital environments, where consumer scrutiny is particularly high.

Fourth, the ethical and regulatory dimension of green advertising demands continued attention. As regulations evolve and awareness of greenwashing intensifies, future research could analyze the effectiveness of existing regulatory frameworks in preventing deceptive practices, the interpretation and enforcement of these regulations by companies, and consumer perceptions of the clarity and stringency of regulations. It would also be relevant to explore the role of self-regulation and industry initiatives in promoting ethical green advertising.

Finally, an emerging line of research could focus on consumer segmentation and personalization of green advertising messages. Given the diversity of environmental awareness levels and values among consumers, future studies could investigate how companies can tailor their messages to resonate more effectively with different audience segments. This could include analyzing the types of arguments (rational vs. emotional), value frames (altruistic vs. selfish), and communication channels preferred by different groups of consumers interested in sustainability.

## SOLUTIONS AND RECOMMENDATIONS

The analysis conducted on green advertising, which has explored its characteristics, campaigns, strategies, effectiveness and ethical challenges, allows us to derive a set of solutions and recommendations aimed at enhancing its positive impact and mitigating the risks inherent to greenwashing. These proposals are aimed at companies, regulators and consumers, seeking to foster more authentic and effective sustainability communication.

First, radical transparency emerges as a fundamental solution for companies. As Peattie (2010) has pointed out, credibility is the cornerstone of sustainable marketing. Companies must go beyond generic claims and provide detailed, verifiable and easily accessible information about their environmental practices. This involves the disclosure of specific data on the product life cycle, the origin of materials, production processes and the impact of their supply chain. Implementing technologies such as blockchain for traceability could be an innovative solution to certify the authenticity of claims, allowing consumers to access information with a high degree of confidence.

A second crucial recommendation is strategic alignment and 360-degree coherence. Green advertising cannot be an isolated marketing strategy; it must be a genuine reflection of a deep organizational commitment to sustainability. This means that green marketing policies must be intrinsically linked to overall corporate strategy, operations, research and development, and organizational culture (Kotler & Keller, 2016). When the advertising message is decoupled from actual business actions, the risk of greenwashing rises exponentially, eroding consumer trust and long-term brand reputation. Companies must invest in the sustainability of their processes and products before communicating it extensively.

Third, for regulators and watchdogs, the creation of clearer, more specific and internationally harmonized regulatory frameworks is imperative. Ambiguity in the definition of terms such as "green" or "sustainable" facilitates deceptive practices. Strict guidelines on the use of labels, certifications and environmental claims, as well as proportionate verification mechanisms and penalties for greenwashing, need to be established. Collaboration between different countries and regions is essential to avoid regulatory loopholes that companies could exploit. Initiatives such as those proposed by the European Commission on green claims are a step in the right direction, but require rigorous implementation.

A fourth solution focuses on consumer education and the development of environmental literacy. An informed consumer is a more critical consumer and less susceptible to greenwashing. Consumer organizations and governments themselves have a vital role to play in awareness campaigns that empower individuals with tools to identify misleading claims and assess authenticity. This includes promoting critical thinking about advertising messages and understanding the true environmental impacts of products and services.

Finally, fostering multi-sector collaboration is a cross-cutting recommendation. Companies, regulators, NGOs and academia should work together to establish best practices, develop standardized sustainability metrics and promote innovation in green communication. Participation in initiatives such as "1% for the Planet," adopted by Patagonia, is an example of how collaboration can lend credibility and amplify impact. Academic research, meanwhile, should continue to explore the complexities of consumer perception and the real impact of green advertising, providing an empirical basis for the development of more effective future strategies and policies.

## CONCLUSIONS AND DISCUSSION

The present research work has sought to provide a comprehensive understanding of the contemporary landscape of green or sustainable advertising, addressing its characteristics, trends, predominant strategies, its impact on brand image and consumer loyalty, as well as the ethical and regulatory challenges inherent to the practice of greenwashing. The descriptive analysis of various campaigns and theoretical reflections derived from the academic literature converge in a series of fundamental conclusions that delineate the complexity and relevance of this field in the current context of growing environmental awareness.

First, the description of the characteristics and trends of green advertising reveals a significant evolution since its origins. Initially marked by simplistic and often lacking in rigorous claims, today's green advertising tends towards greater sophistication, seeking to integrate sustainability more deeply into the brand narrative. There is a growing prevalence of messages emphasizing transparency, supply chain traceability, circular economy and commitment to broader sustainable development goals. The trend towards digitalization has driven the use of new formats and interactive channels, enabling more direct and segmented communication with specific audiences.

Analysis of major green advertising campaigns since 2020 evidences a diversity of strategic approaches. Companies such as Patagonia continue to lead with campaigns focused on authenticity and environmental activism, while others, such as Tesla, base their communication on the inherently sustainable technological innovation of their products. Consumer giants such as Unilever and Coca-Cola implement large-scale campaigns that address specific challenges such as packaging waste and responsible sourcing. IKEA and H&M seek to integrate sustainability into the value proposition of their home and fashion products, respectively. While the themes and tone vary, there is convergence on the need to communicate concrete actions and avoid vague or unsubstantiated claims.

Identifying and categorizing the most commonly employed green advertising strategies reveals a range of tactics. Emphasis on "green" product attributes (recycled materials, energy efficiency) remains a central strategy. However, there is an increase in communicating long-term corporate commitment, promotion of the circular economy (recycling, reuse) and collaboration with environmental organizations. The narrative of transparency and authenticity emerges as a key differentiating factor, seeking to build consumer trust in an environment where skepticism towards greenwashing is growing.

Evaluation of the effectiveness of green advertising in building brand image and consumer loyalty yields nuanced results. Authentic and well-executed green advertising has the potential to strengthen brand image by associating it with positive values of environmental responsibility and commitment. This can generate greater loyalty from environmentally conscious consumers. However, perceived greenwashing can have the opposite effect, eroding trust and damaging brand reputation in the long term. Effectiveness appears to be intrinsically linked to consistency between advertising messages and actual company actions, as well as to the clarity and verifiability of environmental claims.

Finally, analysis of the ethical and regulatory challenges associated with greenwashing underscores the fine line that companies must navigate when communicating their environmental credentials. The temptation to exaggerate benefits or to focus on marginal aspects without addressing significant impacts represents a considerable ethical risk. From a regulatory perspective, a number of global initiatives seek to establish frameworks that prevent misleading advertising and promote transparency. The effectiveness of these regulations and their enforcement vary, but the trend points toward increased scrutiny of environmental claims and the imposition of penalties in cases of proven greenwashing.

Regarding the discussion following this research, we note that we have discovered the complexity and multiple dimensions of a crucial field at the intersection of marketing, communication and sustainability. The conclusions derived from the analysis of campaigns and theoretical reflection raise a rich discussion around the tensions inherent in the promotion of responsible environmental practices, the effectiveness of the strategies employed and the ethical and regulatory challenges that persist.

One of the central points of discussion lies in the dichotomy between the need to communicate sustainability efforts and the latent risk of greenwashing. While companies recognize the growing demand for products and services with environmental credentials, the temptation to present a "greener" image than operational reality supports remains a palpable concern. The proliferation of ambiguous terms, the focus on minor environmental attributes and the lack of transparency in the value chain erode consumer confidence and undermine the credibility of green advertising as a whole. This tension calls for an ongoing debate on ethical standards and the need for more robust regulatory mechanisms to ensure the veracity of environmental claims.

In relation to the effectiveness of green advertising strategies, research suggests that the mere mention of "green" attributes is not sufficient to build a strong brand image and foster consumer loyalty. Authenticity, consistency between message and action, and the ability to connect with consumers' intrinsic values of sustainability emerge as key factors. Campaigns that transcend superficial product promotion and engage in narratives of purpose, transparency and long-term commitment appear to generate deeper and more lasting resonance. However, accurately measuring the impact of green advertising on loyalty and purchase behavior remains a methodological challenge that requires the integration of data from various sources and analytical approaches.

The influence of industry context and business model on green advertising strategies also deserves particular attention in the discussion. The companies analyzed, belonging to diverse industries, demonstrate approaches tailored to their specific sustainability challenges and opportunities. While Patagonia integrates environmental awareness into the core of its brand identity, companies like Tesla capitalize on technological innovation as an inherently "green" solution. FMCG giants such as Unilever and Coca-Cola face the complexity of addressing sustainability in extended value chains and the need to communicate messages to broad and heterogeneous audiences. These sectoral differences highlight the absence of a "one size fits all" in green advertising and the importance of developing contextually relevant strategies. The growing relevance of digital media and interactive platforms opens up new avenues for sustainability communication, but also poses challenges in terms of information verification and credibility management. The immediacy and amplification capabilities of social media can be both a powerful tool for disseminating authentic messages and a breeding ground for misinformation and greenwashing. Future discussion must address how companies can leverage these channels in an ethical and transparent manner, encouraging consumer engagement and building communities around shared sustainability values.

Finally, the interplay between ethics and regulation emerges as a central focus of the discussion. While business ethics demand honest and responsible communication, the existence of regulatory gaps and the difficulty of standardizing and verifying environmental claims create fertile ground for greenwashing. Increasing consumer pressure and the action of international regulatory bodies point to a trend towards greater scrutiny and the need for clearer and more effective regulatory frameworks. The discussion should focus on how to balance the legitimate promotion of sustainability efforts with consumer protection against misleading practices, fostering a culture of transparency and accountability in green advertising.

## REFERENCES


Banerjee, S. B. (2020). *Corporate social responsibility: The good, the bad, and the ugly*. Edward Elgar Publishing.

Brands Called Out For Greenwashing In 2020 - CWR - China Water Risk. (2020, diciembre 17). *China Water Risk*. Retrieved from https://cwrrr.org/opinions/8-brands-called-out-for-greenwashing-in-2020/

Crane, A. (2000). Facing the backlash: Green marketing and strategic ambiguity. *Journal of Strategic Marketing*, *8*(1), 1–17.

Dahlén, M., Lange, F., & Smith, T. (2010). *Marketing communications: A brand narrative approach*. John Wiley & Sons.

Font, X., & Elgazzar, S., &лом, J. (2020). Greenwashing and beyond: The need for clear, transparent and evidence-based communication about tourism sustainability. *Journal of Sustainable Tourism*, *28*(7), 959–977.

Gbadamosi, A. (2019). Marketing ethics, green and sustainable marketing. *Contemporary Issues in Marketing: Principles and Practice*, 185-213.

Kilbourne, W. E. (2004). Sustainable marketing: The new marketing paradigm for a new millennium. *Marketing Intelligence & Planning*, *22*(7), 702–710.

Kotler, P., & Keller, K. L. (2016). *Marketing management* (15th ed.). Pearson.

Leonidou, L. C., (2020). “Green” advertising effects on consumer responses: A contingency perspective. *Journal of Advertising*, *49*(1), 1–20.

Lyon, T. P., & Montgomery, A. W. (2015). The means and end of greenwash. *Organization & Environment*, *28*(2), 223–249. DOI: 10.1177/1086026615575332

Meadows, D. H., Meadows, D. L., Randers, J., & Behrens, W. W.III. (1972). *The limits to growth*. Universe Books.

Ottman, J. A. (2011). *The new rules of green marketing: Strategies, tools, and inspiration for sustainable branding*. Greenleaf Publishing.

Peattie, K. (2010). Green marketing: Legend, myth, delusion, or panacea? *Journal of Strategic Marketing*, *18*(2), 109–117.

Prieto-Castillo, L., Mijarra, J. J., & Rubio-Romero, J. C. (2021). La publicidad verde como herramienta para el consumo responsable: Un análisis de la percepción del consumidor. *Revista Mediterránea de Comunicación*, *12*(1), 145–159.

# Chapter 5
# Greenwashing as a Corporate Communication Strategy in Tourism Companies

**José Ramón Sarmiento-Guede**
https://orcid.org/0000-0002-0342-0348
*Universidad Rey Juan Carlos, Spain*

**Alberto Azuara-Grande**
https://orcid.org/0000-0002-8432-0065
*Universidad Rey Juan Carlos, Spain*

## ABSTRACT

*In Smart Cities, tourism companies increasingly promote sustainability, but many engage in greenwashing, misleading communication that exaggerates environmental responsibility. This chapter analyzes how such practices affect consumer trust and purchasing decisions. Using a Case Study approach, this chapter reveals that greenwashing generates skepticism and harms corporate reputation, while transparency and verifiable sustainability claims foster consumer confidence. The chapter offers practical recommendations for ethical communication, highlights the role of regulation and consumer education, and suggests future research on loyalty and technological tools to verify sustainability efforts.*

DOI: 10.4018/979-8-3373-2802-7.ch005

## INTRODUCTION

In recent decades, corporate communication has evolved significantly, becoming a key tool for managing reputation and stakeholder relationships. In the tourism sector, where sustainability has become a fundamental value, companies have implemented communication strategies aimed at demonstrating their commitment to the environment and sustainable development (Font & McCabe, 2017). However, in many cases, these strategies do not reflect actual actions but rather respond to a misleading practice known as greenwashing (Delmas & Burbano, 2011).

The growing environmental awareness among consumers has generated a demand for responsible tourism services, prompting companies to adapt their communication to highlight their commitment to sustainability (Dodds, Holmes, & Novotny, 2022). Nevertheless, some organizations have chosen to exaggerate or distort their environmental initiatives to improve their corporate image without making significant changes to their operations. This phenomenon of greenwashing presents both an ethical and economic dilemma, as it not only deceives consumers but can also generate distrust and harm the reputation of the tourism sector (Lyon & Montgomery, 2015).

Greenwashing is defined as the practice of presenting a false or exaggerated image of sustainability through communication strategies that do not correspond to actual actions (Bowen, 2014). In the tourism context, this strategy can manifest through weak environmental certifications, misleading advertising, or ambiguous claims about the ecological impact of their operations. A common example is the promotion of hotels that label themselves as "eco-friendly" without meeting strict sustainability standards, which confuses consumers and undermines genuine responsible tourism initiatives (Torelli, Balluchi & Lazzini, 2020).

From an economic perspective, greenwashing can generate short-term benefits by attracting customers who value sustainability. However, in the long run, it can damage the credibility of both the company and the sector in general. Ethically, this practice is questionable because it exploits consumers' goodwill and undermines trust in truly ecological initiatives. Moreover, greenwashing can hinder progress toward truly sustainable tourism by diverting attention and resources away from practices that generate real positive environmental impacts (Peattie & Crane, 2005).

The importance of this chapter lies in the need to analyze how tourism companies use greenwashing in their corporate communication and how this strategy affects consumers and the sector. Research in this area allows for the identification of greenwashing patterns and the development of strategies to foster more transparent and ethical communication in tourism (Font, Elgammal & Lamond, 2017). Through the analysis of practical cases, this chapter seeks to highlight the consequences of greenwashing and propose recommendations for more responsible corporate management in the tourism sector.

Greenwashing represents a challenge for corporate communication in tourism companies, as it affects consumer trust and the credibility of sustainability initiatives. Therefore, it is essential to understand its mechanisms and consequences to promote more ethical and effective practices in the management of tourism sustainability.

## RESEARCH QUESTIONS AND OBJECTIVES

Greenwashing, understood as the practice of companies exaggerating or falsifying their environmental initiatives to enhance their image (Delmas & Burbano, 2011), has become a growing concern in the tourism sector. As consumers seek more sustainable options, some companies adopt corporate communication strategies that may mislead about their environmental commitment (Leonidou & Skarmeas, 2017). This raises questions about how this practice is presented, its effects on consumer perception, and the mechanisms to identify and prevent it. Based on these challenges, the following research questions and objectives have been formulated:

- RQ1. How do tourism companies use greenwashing in their corporate communication?
- RQ2. What are the impacts of greenwashing on consumer perception?
- RQ3. What strategies can help identify and prevent greenwashing in the tourism sector?

To answer these questions, we propose the following objectives:

- RO1. Analyze the corporate communication strategies related to greenwashing in tourism companies.
- RO2. Identify concrete cases in which this practice has been evidenced.
- RO3. Evaluate the impact of greenwashing on consumer trust and corporate reputation.
- RO4. Propose more transparent and ethical communication strategies for sustainable tourism.

## THEORETICAL FRAMEWORK

The term "greenwashing" was coined by Jay Westerveld in 1986 to describe the practice of companies exaggerating or falsifying their environmental credentials to improve their corporate image (Westerveld, 1986, in Delmas & Burbano, 2011). Since then, the concept has evolved and has been analyzed by various authors. Delmas and Burbano (2011) define greenwashing as the misleading communication by a company about its environmental practices, which can lead to consumer confusion and undermine trust in corporate sustainability.

According to Lyon and Montgomery (2015), greenwashing can manifest through advertising messages, product labeling, and sustainability reports that lack independent verification. In the tourism sector, Font and Lynes (2018) highlight how some companies promote weak environmental initiatives for marketing purposes, without implementing real changes in their operations.

*Table 1. Definitions of greenwashing*

| Author and Date | Definition of Greenwashing |
|---|---|
| Laufer, W. S. (2003) | "Greenwashing is the use of communication and marketing strategies by companies to project a misleading image of environmental responsibility without implementing actual sustainable practices." |
| Delmas, M. A., & Burbano, V. C. (2011) | "Greenwashing refers to misleading communication by an organization about its environmental practices to create a positive image that does not reflect reality." |
| Parguel, B., Benoît-Moreau, F., & Larceneux, F. (2011) | "Greenwashing is a marketing strategy where companies use ambiguous or deceptive environmental messages to enhance their image without making substantial changes in their operations." |
| Marquis, C., & Toffel, M. W. (2012) | "Greenwashing occurs when a company invests more in communicating its environmental practices than in actually implementing them." |
| Walker, K., & Wan, F. (2012) | "Greenwashing refers to the discrepancy between a company's positive environmental communication and its actual environmental performance." |
| Bowen, F., & Aragon-Correa, J. A. (2014) | "Greenwashing is the misleading communication of an organization's environmental practices to gain reputational benefits without a genuine commitment to sustainability." |
| Nyilasy, G., Gangadharbatla, H., & Paladino, A. (2014) | "Greenwashing is the practice of making misleading or exaggerated environmental claims to create an impression of environmental responsibility." |
| Leonidou, C. N., & Skarmeas, D. (2015) | "Greenwashing involves the adoption of misleading environmental communications to create a favorable image without real environmental performance to support it." |

**Source:** Authors

## Corporate Communication in the Tourism Sector

Corporate communication plays a key role in the perception of sustainability in tourism. According to Morsing and Schultz (2006), an effective communication strategy must be based on transparency and the verifiability of environmental claims.

In the context of sustainable tourism, Morsing and Schultz (2006) suggest that companies should avoid exaggerating their green credentials and focus on tangible actions to reduce their environmental footprint.

Green marketing is a tool used by tourism companies to attract environmentally conscious consumers (Leonidou & Skarmeas, 2017). However, if genuine sustainable practices are not implemented, these strategies may be perceived as misleading, generating skepticism among consumers (Rahman & Nguyen-Viet, 2023).

In the tourism field, sustainable corporate communication must be clear, verified, and aligned with concrete actions. A company's credibility largely depends on its ability to provide accurate and verifiable information about its environmental efforts (Font & Lynes, 2018). For this reason, many companies have adopted sustainability reports based on international standards such as those of the Global Reporting Initiative (GRI) or the Carbon Disclosure Project (CDP), which help support their environmental claims (Camilleri, 2022).

A key aspect of communication in sustainable tourism is stakeholder engagement. According to Morsing and Schultz (2008), companies must actively involve consumers, employees, suppliers, and local communities in their ecological initiatives to build trust and engagement. Two-way dialogue strategies, such as satisfaction surveys, social media, and discussion forums, can improve public perception of the authenticity of a company's environmental efforts (Morsing & Schultz, 2008).

On the other hand, sustainable communication in tourism faces several challenges. One of them is the difficulty in distinguishing between authentic green marketing and greenwashing strategies. Research has shown that consumers are increasingly skeptical of companies' environmental claims, especially when they are not supported by recognized certifications (Chen & Chang, 2013). Furthermore, the proliferation of eco-labels without clear regulation can contribute to consumer confusion and weaken trust in the sector (Gössling & Buckley, 2016).

To counter these challenges, tourism companies must adopt principles of responsible communication. According to Goodman and Salleh (2013), transparency is an essential factor in consumer perception and can be achieved through the disclosure of verifiable data, independent audits, and third-party involvement in the evaluation of sustainable practices. Moreover, the narrative used in environmental communication must be consistent with the company's actual actions, avoiding exaggerated or ambiguous messages that may be interpreted as misleading (Rahman et al., 2023).

Finally, it is important to highlight the role of technology in sustainable corporate communication. Digital tools such as mobile apps, interactive websites, and social media allow companies to provide detailed, real-time information about their ecological initiatives. Some hotels and airlines have implemented platforms where customers can track the carbon footprint of their trips and learn about the actions the company is taking to mitigate its environmental impact (Gössling & Peeters, 2015).

## Marketing Strategies and Corporate Sustainability

Sustainable marketing is defined as a set of business strategies that integrate environmental, social, and economic sustainability into a company's value proposition (Belz & Peattie, 2012). Its main goal is to promote products and services that minimize negative environmental impact and contribute to the preservation of natural and cultural resources. In the tourism sector, sustainable marketing is especially relevant due to travelers' growing environmental awareness and their preference for experiences that respect the environment (Font & McCabe, 2017). Marketing strategies that promote sustainability may focus on various aspects, such as:

- Responsible tourism: Promoting destinations and activities that respect biodiversity, minimize carbon footprints, and benefit local communities (Goodwin, 2011).
- Ecological certifications: Obtaining seals and accreditations that ensure compliance with environmental standards, such as the Green Globe certification or the Blue Flag label (Buckley, 2012).
- Transparency in environmental management: Communicating verifiable data on emission reduction, use of renewable energy, and recycling policies implemented by hotels and tour operators (Dodds & Holmes, 2011).
- Social commitment: Initiatives that promote equity, fair trade, and the development of local communities through tourism (Scheyvens, 2011).

However, for these strategies to be effective, they must be backed by concrete actions and not merely consist of discourse. A lack of consistency between what is communicated and what is actually implemented can lead to greenwashing, damaging the company's credibility and reputation (Delmas & Burbano, 2011). Over the years, various tourism companies have successfully integrated sustainable marketing strategies, achieving both business benefits and positive environmental impacts. Some best practices include:

- Transparency-based strategies: Companies like Patagonia and National Geographic Expeditions have implemented environmental transparency policies, openly detailing their ecological impacts and the measures taken to mitigate them. In tourism, hotel chains such as Iberostar have published their plans for reducing plastics and energy consumption, demonstrating real commitment to sustainability (Dwyer, 2018).
- Collaboration with certifying bodies: Obtaining environmental certifications from independent organizations lends credibility to sustainability strategies. For example, the EarthCheck certification has been adopted by several hotel

chains and tourist destinations to validate their sustainability practices and distinguish them from greenwashing (Honey, 2008).
- Use of renewable energy and waste reduction: Some tourism companies have incorporated solar energy, wind turbines, and advanced recycling systems into their operations. A notable example is the Six Senses group, which operates luxury hotels with a sustainable approach, minimizing the consumption of natural resources and promoting environmental conservation (Gössling & Hall, 2019).
- Promotion of community-based tourism: Companies like G Adventures and Intrepid Travel have focused their marketing strategies on promoting authentic tourism experiences that generate economic benefits for local communities. This approach not only contributes to social and economic sustainability but also enhances the travel experience by offering direct contact with local culture (Ashley & Mitchell, 2009).

Despite the growing interest in sustainability, many companies have used the concept opportunistically, resorting to greenwashing as a deceptive marketing strategy. Some of the most common greenwashing practices include (Parguel, Benoît-Moreau & Larceneux, 2011):

- Use of vague and unsubstantiated terms: Expressions such as "eco-friendly", "sustainable", or "green" without concrete data to support such claims.
- Exaggeration of ecological initiatives: Highlighting a single isolated action as if it represented a complete transformation toward sustainability.
- Misleading images and symbols: Use of green colors, leaf icons, and natural landscapes without a genuine commitment to sustainability.
- Poorly verifiable carbon offsets: Companies claiming to offset their carbon footprint without offering clear proof of how it is done.

The problem with greenwashing is that it not only misleads consumers but also harms companies that have genuinely made efforts to reduce their environmental impact. In response to this issue, several solutions have been proposed to ensure more ethical communication and avoid misinformation in tourism marketing. To avoid greenwashing and ensure that sustainable marketing strategies are perceived as credible, companies must adopt principles of transparency and verification of their environmental initiatives. Key recommendations include (Font & McCabe, 2018):

- Providing verifiable data: Presenting statistics and concrete evidence on the sustainable actions implemented, such as $CO_2$ emission reductions, recycling rates, or renewable energy consumption.

- Using recognized certifications: Opting for environmental certifications from independent bodies, such as the LEED certification for sustainable hotels or the Rainforest Alliance seal for tour operators.
- Avoiding ambiguous terms and generalizations: Rather than claiming a hotel is "100% green", it is preferable to detail the specific actions that make it sustainable (Carlson, Grove & Kangun, 1993).
- Involving consumers: Allowing tourists to participate in environmental initiatives, such as wildlife conservation programs, plastic reduction, or reforestation (Boley & Uysal, 2013).
- Transparency in sustainability reporting: Publishing annual reports on the company's environmental and social impact, including external audits, to support the data.

## Consumer Perception of Greenwashing

Consumer perception plays a crucial role in the effectiveness of sustainable marketing and the credibility of ecological initiatives by tourism companies. According to Chen and Chang (2013), when consumers identify greenwashing practices, their trust in the brand decreases significantly, which impacts both their purchasing intention and brand loyalty.

Today's consumers are increasingly informed and demand concrete proof of a company's environmental commitment before making purchasing decisions. Research such as Souza (2024) shows that travelers actively seek recognized environmental certifications, third-party assessments, and tangible evidence that a company's sustainable practices are genuine. This means that companies that fail to support their environmental claims with verifiable actions may lose credibility in the market.

In the hospitality industry, Millar and Baloglu (2011) concluded that consumers prefer establishments with recognized certifications rather than those that merely make unsupported claims. Moreover, tourists view greenwashing practices negatively when they discover that a company's environmental efforts are superficial or inconsistent.

Another relevant aspect is the impact of electronic word of mouth (eWOM) on consumer perception. Social media and review platforms allow travelers to share their experiences regarding the sustainability of tourism establishments. A study by Rahman and Nguyen-Viet (2023) shows that tourism companies accused of greenwashing can suffer significant reputational damage, leading to reduced demand and financial loss.

To avoid negative consumer perception, companies should focus on transparency and verifiability in their sustainable practices. Honest communication backed by verifiable data is essential to build trust and strengthen customer relationships.

## Impact of Greenwashing on Corporate Reputation

In a business environment where sustainability has become a key factor for differentiation and competitiveness, many companies have turned to greenwashing as a strategy to attract environmentally conscious consumers (Delmas & Burbano, 2011). However, this practice carries significant risks for corporate reputation. When consumers detect inconsistencies between a company's environmental rhetoric and its actual actions, their trust declines, which may lead to a brand image crisis and a loss of customer loyalty (Parguel, Benoît-Moreau & Larceneux, 2011).

Corporate reputation is an intangible asset that influences how consumers perceive and trust an organization. In the context of tourism—where experience and credibility are essential—reputation can determine customers' preference for certain companies or destinations (Pulido-Fernández & López-Sánchez, 2016). When a company engages in greenwashing, its reputation can be compromised in various ways:

- Loss of consumer trust: Today's consumers are increasingly critical and better informed about corporate practices. If they perceive that a company has exaggerated or falsified its environmental commitment, they may react with skepticism and rejection.
- Impact on brand loyalty: Consumer loyalty is built on credibility and authenticity. When a company is caught greenwashing, its perceived reliability drops, which can lead to a decrease in customer retention.
- Deterioration of corporate image: A company's image is key to its market positioning. A greenwashing scandal can lead to negative media coverage and damage its long-term reputation.
- Financial impact: Declines in consumer trust and corporate reputation can result in economic losses. Studies have shown that companies perceived negatively in terms of sustainability experience a drop in sales and share value.

In the digital age, social media plays a fundamental role in building and damaging corporate reputation. Platforms like Twitter (X), Facebook, and Instagram allow consumers to report and expose greenwashing practices in real time, amplifying the negative impact for involved companies (Leonidou, Skarmeas & Saridakis, 2018). A relevant example is the #Greenwashing movement, which has gained popularity on social media, where users share examples of misleading sustainability advertising. This phenomenon has forced many companies to be more careful with their environmental communication, as greenwashing exposure can go viral and spark a reputational crisis in a matter of hours (Dahl, 2010).

The impact of greenwashing on corporate reputation is not only immediate but can also have long-term repercussions. Key consequences include:

- Difficulty regaining trust: Once a company's credibility is damaged, restoring consumer trust is challenging. In many cases, companies must invest heavily in communication campaigns to repair their image (Delmas, Lyon & Maxwell, 2019).
- Increased scrutiny by regulators: Companies flagged for greenwashing may face tighter monitoring by environmental regulatory bodies, limiting their flexibility in future marketing campaigns (Lyon & Montgomery, 2015).
- Loss of customers and strategic partnerships: In a market where consumers increasingly value sustainability, companies with a history of greenwashing may lose clients and miss out on collaboration opportunities with business partners seeking ethical and trustworthy brands (Kahneman & Tversky, 1979)

## Regulations and Certifications in Sustainable Tourism

To prevent greenwashing and ensure sustainability in the tourism sector, various regulations and certifications aim to establish clear and verifiable standards. At the international level, the World Tourism Organization (UNWTO) has developed sustainable tourism principles that emphasize the need for responsible management of natural and cultural resources (UNWTO, 2019). Additionally, the Kyoto Protocol and the Paris Agreement set global commitments to reduce carbon emissions, which directly impact the tourism industry (United Nations, 2015). These regulations encourage companies to adopt more sustainable practices and mitigate their environmental impact. In terms of certifications, there are internationally recognized labels that guarantee the implementation of verified sustainable practices in hotels and other tourism services. Some of the most notable certifications include:

- Green Key: An award given to tourist accommodations and restaurants that meet strict environmental criteria, such as efficient resource management and waste reduction.
- EarthCheck: A certification program based on assessing the environmental performance of tourism businesses, including energy use, water management, and waste handling.
- LEED (Leadership in Energy and Environmental Design): A certification system for sustainable buildings, covering the efficient design, construction, and operation of resources.
- Biosphere Tourism: A certification promoted by the Responsible Tourism Institute (RTI), endorsing the comprehensive sustainability of destinations and tourism companies according to the UN's Sustainable Development Goals (SDGs).

Despite the existence of these certifications, studies have shown that some companies obtain these labels without implementing significant operational changes, contributing to consumer distrust (Gössling & Buckley, 2016). Furthermore, the proliferation of eco-labels without clear regulation can create confusion among consumers and weaken the credibility of the sustainable tourism sector.

To prevent the misleading use of certifications, it is recommended that they be granted by independent entities with strict evaluation criteria. Additionally, regular verification and external audits can ensure that certified companies continue to meet the required standards over time.

## METHODOLOGY

This study adopts a qualitative approach based on case analysis, a methodology that enables an in-depth examination of the greenwashing phenomenon in the tourism sector. According to Yin (2018), case study research is an ideal strategy for analyzing complex phenomena in their real-life context, which is particularly relevant for evaluating the impact of corporate communication strategies on consumer perception and corporate reputation.

To select the cases, criteria were established to ensure the relevance and validity of the analysis. First, tourism companies with international presence were considered, as their global reach allows for an assessment of how greenwashing practices may manifest in different markets (Font, Elgammal & Lamond, 2017). Likewise, the study includes companies that have been criticized for allegedly deceptive sustainability practices, as well as those recognized for their transparency and genuine environmental commitment. Comparing these cases helps better understand differences in corporate communication and their impact on public opinion (Delmas & Burbano, 2011).

The research draws on various secondary information sources to enable a thorough analysis of corporate discourse and its alignment with actual practices. Sustainability reports published by the selected companies were examined, as these documents serve as primary sources to understand the environmental actions and commitments claimed by the companies (Gössling & Buckley, 2016). In addition, corporate communication campaigns were reviewed—including advertisements, social media content, and websites—to identify discursive strategies used to project a sustainable image (Rahman, Park & Chi, 2017). Academic studies, reports from international organizations, and expert opinions on sustainable tourism were also considered to contextualize findings and assess the credibility of the analyzed practices.

The data analysis was developed using two main strategies. First, corporate discourse was compared with the actual actions of each company, identifying potential discrepancies between what is communicated and what is truly implemented. This contrast is essential to detect greenwashing practices and evaluate the authenticity of the sustainability initiatives promoted by the companies (Millar & Baloglu, 2011). Second, the impact of these strategies on corporate reputation and consumer perception was evaluated. To do this, previous studies on customer trust in sustainable brands were considered, along with analyses of reactions on digital platforms and specialized forums (Leonidou & Skarmeas, 2017).

This methodological approach enables a comprehensive understanding of the greenwashing issue in tourism, providing tools to distinguish between genuine sustainability strategies and those that seek to manipulate consumer perception. Moreover, the findings can contribute to formulating recommendations for more ethical and transparent corporate communication in the sector.

## CASE STUDY DEVELOPMENT

Sustainability has become a key factor in the tourism sector today. Hotel companies, airlines, and booking platforms have adopted marketing strategies based on environmental commitments to attract an increasingly eco-conscious public. However, in many cases, these initiatives have been criticized due to a lack of transparency, failure to fulfill commitments, and the use of misleading communication strategies. The phenomenon of greenwashing has gained relevance in the tourism industry, as many companies promote sustainability policies that, in practice, have little impact or lack verification. Such practices not only affect the credibility of the companies involved but also generate consumer distrust and hinder the advancement of truly sustainable solutions in the industry. The following section analyzes five practical cases of tourism companies that have been accused of greenwashing or, conversely, have implemented transparent communication strategies and genuine sustainability efforts. Through the study of Accor Hotels, Ecocamp Patagonia, Marriott International, Ryanair, and Booking.com, we will examine the differences in their sustainability approaches, public responses, and the consequences of their strategies.

### Case 1: Accor Hotels and Greenwashing Accusations

Accor Hotels, one of the largest hotel chains in the world, has faced criticism over its sustainability strategies. In 2021, the company announced its “Planet 21” initiative, promising to reduce its carbon footprint and promote sustainable practices in its hotels (Accor, 2021). However, independent reports and environmental organizations have questioned the authenticity of these claims, accusing the company of engaging in greenwashing.

Accor's strategy relied heavily on a marketing campaign focused on eliminating single-use plastics and promoting hotels with eco-certifications. However, a detailed analysis of its internal operations revealed contradictions. For example, while the company promoted the elimination of plastic straws, it continued to use plastic packaging in minibars and toiletries, indicating a partial implementation of its commitments.

Consumers and environmental activists reacted skeptically to these measures, especially when it was revealed that Accor continued to expand its ecological footprint by building new hotels in environmentally sensitive areas. According to a Greenpeace report (2022), Accor's sustainability marketing was not backed by a significant reduction in the company's carbon emissions, weakening its market credibility.

This situation had reputational consequences for the company, affecting its relationship with clients and business partners. A study by Brown and Smith (2023) notes that companies accused of greenwashing tend to experience a 15% decline in customer purchase intent, highlighting the importance of aligning discourse with action.

## Case 2: Ecocamp Patagonia and its Transparent Communication Strategy

In contrast to the previous case, Ecocamp Patagonia has been recognized as a model of sustainable tourism due to its transparent and action-based approach. Located in Torres del Paine National Park, Chile, this hotel is known for its use of renewable energy, waste treatment, and low-impact construction.

Unlike Accor, Ecocamp Patagonia has adopted a communication strategy based on verifying its environmental commitments through independent certifications such as "B Corp" and "Leave No Trace." Moreover, the company not only presents its achievements but also communicates its challenges and limitations, fostering a trusting relationship with its customers.

A study by Jones et al. (2023) indicates that consumers perceive companies that openly acknowledge their areas for improvement as more credible, reinforcing the authenticity of their sustainability commitment.

The impact of this strategy has been highly positive. In 2022, Ecocamp Patagonia was recognized with the "World Travel Award" in the sustainable tourism category, boosting its appeal among environmentally conscious travelers. According to a report by Sustainable Travel International (2023), 72% of tourists prefer to stay at establishments with verifiable sustainability practices, reinforcing the company's transparency-based strategy.

## Case 3: Marriott International and its "Serve 360" Program

Marriott International, one of the largest hotel chains in the world, has widely promoted its sustainability and social responsibility program "Serve 360: Doing Good in Every Direction." This program focuses on four core pillars: environmental sustainability, community engagement, diversity and inclusion, and professional development. However, despite these ambitious claims, multiple reports have indicated that the company has engaged in greenwashing, as many of its initiatives have not translated into real operational changes.

One of the main points of criticism toward Marriott has been its policy on reducing single-use plastics. In 2019, the company announced the elimination of individual-size shampoo and conditioner bottles across its hotels, replacing them with reusable dispensers. While this measure was presented as a significant effort to reduce plastic waste, environmental experts have noted that such changes have minimal impact compared to other pollution sources in the hotel industry, such as water and energy waste in large resorts.

Moreover, reports have revealed that Marriott continues to operate hotels in environmentally sensitive areas and promotes luxury tourism without implementing effective measures to mitigate its ecological impact. For example, the chain's expansion into fragile destinations such as the Caribbean and Southeast Asia has raised concerns among environmentalists due to the destruction of coastal ecosystems and increased greenhouse gas emissions associated with mass tourism. A study by Rainforest Alliance (2022) indicated that many Marriott properties failed to meet the minimum sustainability standards required by international certifications, despite promoting themselves as "eco-hotels."

Another point of controversy has been carbon offsetting. Marriott has claimed to work on neutralizing its carbon footprint through investments in reforestation and renewable energy projects. However, investigations by Carbon Market Watch (2023) have shown that several of these projects lack transparency and do not adequately offset the actual impact of the hotel chain's operations.

As a result, the "Serve 360" program has been perceived by some sectors as a marketing strategy rather than a genuine commitment to sustainability. Consumer organizations have warned that companies engaging in greenwashing may suffer significant reputational consequences. A study by Jones and Patel (2023) found that customers are increasingly skeptical of unverified sustainability claims, which has impacted trust in the Marriott brand.

## Case 4: Ryanair and its Low Emissions Advertising

Ryanair, one of Europe's largest low-cost airlines, has used advertising campaigns to promote its alleged leadership in carbon emission reduction. In ads broadcast across several European countries, the company claimed to be "Europe's lowest-emission airline," highlighting its fleet's efficiency and sustainability efforts. However, various investigations have shown these claims to be misleading, and Ryanair remains one of the continent's highest carbon emitters.

In 2019, the European Environment Agency listed Ryanair among the top ten carbon emitters in Europe—a category historically dominated by coal-based power companies. This data directly contradicts the company's narrative about its low environmental impact. According to a Transport & Environment report (2021), although Ryanair has improved aircraft efficiency, the continued expansion of its low-cost flight network has led to an absolute increase in emissions, offsetting any technological improvements.

Furthermore, Ryanair's advertising has faced legal scrutiny. In 2020, the UK Advertising Standards Authority (ASA) ordered the removal of several of the airline's ads for being misleading. The ASA ruled that Ryanair failed to provide sufficient evidence to support its claims of being the "greenest airline," and that its ads misled consumers by omitting relevant information about its actual environmental impact.

Another controversial point is the airline's carbon offsetting program. Ryanair offers passengers the option to contribute to environmental projects to "neutralize" flight emissions, but studies have shown these projects have limited impact and do not fully offset the airline's carbon footprint. An analysis by Carbon Market Watch (2022) concluded that airlines relying on offset programs instead of directly reducing emissions are perpetuating an unsustainable business model.

Ryanair's case clearly illustrates how greenwashing can affect public perception of a company. According to a European Commission survey (2023), 70% of consumers believe airlines are not transparent about their environmental impact, reflecting growing distrust in the sector. This skepticism may translate into changes in consumer behavior, with increased preference for more sustainable transport options such as high-speed rail for short routes.

In summary, while Ryanair has attempted to project a sustainable image through its advertising, the evidence suggests that its practices do not fully support these claims. Regulatory actions and growing environmental awareness among consumers may compel the airline to adopt more substantial measures in the future to avoid further greenwashing accusations.

## Case 5: Booking.com and its Sustainable Travel Claims

Booking.com, one of the world's largest booking platforms, has promoted several initiatives to position itself as a key player in sustainable tourism. In its advertising campaigns and website, the company has highlighted its "Sustainable Travel" program, which supposedly helps travelers find accommodation with lower environmental impact. However, several studies have questioned the truthfulness of these claims, accusing the company of greenwashing.

Since 2021, the platform has introduced "Sustainable Travel" labels to thousands of hotels and accommodations, claiming they meet certain ecological standards. According to Booking.com, this label helps consumers make more responsible decisions and encourages the adoption of sustainable practices in the hospitality industry.

Despite its rhetoric, several investigations have shown that many accommodations labeled as "Sustainable Travel" do not meet strict sustainability criteria. A report by NGO Earthsight (2023) found that hotels in the program often lacked verifiable environmental certifications and, in some cases, their practices contradicted sustainability principles, such as wasting water and energy. Experts have also pointed out that Booking.com does not provide detailed information about how accommodation is classified as sustainable, leading to consumer confusion. In 2022, the UK Competition and Markets Authority (CMA) launched an investigation into possible misleading practices in promoting these accommodations.

The Booking.com case reflects growing consumer distrust regarding sustainability promises in the tourism sector. A study by the European Commission (2023) revealed that 65% of travelers believe online booking companies are not transparent about their environmental impact. As a result, more travelers are seeking platforms that provide clear and verifiable information about eco-friendly practices. Although Booking.com has attempted to position itself as a leader in sustainable tourism, the lack of transparency and greenwashing accusations have affected its credibility. This case underscores the importance of stricter regulation in environmental advertising and the need to offer consumers clear, verifiable information. If the company does not take more concrete measures to ensure the authenticity of its claims, it may face increased sanctions and a decline in public trust.

## Comparative Analysis of the Case Studies

Several notable cases illustrate the complexities and challenges surrounding greenwashing practices in the tourism industry. As environmental awareness grows among consumers, companies increasingly face pressure to demonstrate sustainability. However, when corporate initiatives fail to align with genuine environmental improvements, accusations of greenwashing can quickly emerge, eroding trust and prompting regulatory scrutiny.

Accor Hotels, a major player in the hospitality sector, faced criticism for promoting its "Planet 21" sustainability program while making only limited changes in practice. For example, although the company reduced the use of single-use plastics, other environmentally harmful practices persisted, such as inefficient resource use and limited investment in structural sustainability changes. Despite adopting eco-certifications and launching public awareness campaigns, environmental organizations such as Greenpeace questioned the credibility of Accor's efforts. Critics argued that the company prioritized marketing over measurable impact. As a result, consumer trust was undermined, potentially affecting brand loyalty and purchasing behavior.

In contrast, Ecocamp Patagonia, also in the hospitality industry, has not been subject to greenwashing accusations. The company demonstrates genuine sustainability using renewable energy, waste minimization, and verifiable certifications such as B Corp and Leave No Trace. Furthermore, Ecocamp's architecture is designed to blend harmoniously with the natural environment, reflecting a commitment to low-impact tourism. Its efforts have met with positive public reception and recognition through sustainable tourism awards, which have contributed to a strengthened reputation and increased preference among environmentally conscious travelers. This case exemplifies how authentic sustainability can become a competitive advantage.

Marriott International implemented the "Serve 360" program, which has been criticized for being superficial. While the company eliminated single-use plastics and invested in reforestation and renewable energy projects, larger issues—such as water waste, energy consumption, and expansion into ecologically sensitive areas—remained unaddressed. Moreover, many of the environmental benefits cited were difficult to independently verify. Concerns have also been raised regarding the transparency and effectiveness of Marriott's carbon offset initiatives. This has led to growing skepticism and a decline in consumer trust, particularly among stakeholders demanding quantifiable progress rather than symbolic gestures.

Ryanair, operating in the airline sector, was accused of misleading advertising related to low emissions. The airline promoted itself as "Europe's greenest airline," despite increasing overall emissions due to business expansion. Although the company invested in more fuel-efficient aircraft and launched a carbon offset program, critics argue these measures were insufficient to counteract the broader environmental impact of the airline's operations. These actions triggered regulatory investigations by bodies such as the UK's Advertising Standards Authority (ASA), and environmental NGOs criticized the airline's claims. Consequently, Ryanair suffered from negative public perception and a general mistrust toward sustainability claims in the aviation sector, which is already under scrutiny due to its high carbon footprint.

Finally, Booking.com, a leading reservation platform, introduced a "Sustainable Travel" label for accommodations. However, the initiative lacked clear criteria and verifiable certifications, raising concerns about transparency and misleading communication. Many properties labeled as "sustainable" had no third-party validation or visible environmental practices. The UK's Competition and Markets Authority (CMA) opened an investigation into the platform's sustainability claims, highlighting the need for standardization and regulatory oversight. This led to consumer confusion, reputational damage, and increased calls for stricter regulation in the industry. The case illustrates how digital platforms must balance visibility for sustainability initiatives with rigorous accountability.

## CONCLUSIONS

This study has enabled an analysis of the phenomenon of greenwashing in the corporate communication of tourism companies, revealing that this practice is often used as a marketing strategy to project an image of environmental commitment that does not always translate into concrete actions. Through persuasive discourse, some companies highlight isolated ecological initiatives or use ambiguous language that leads consumers to perceive them as environmentally responsible. However, upon analyzing various cases, it has been shown that this tactic can be counterproductive when consumers detect inconsistencies between the discourse and the actual operations of the companies, negatively impacting their reputation and market trust.

The impact of greenwashing on consumer perception is significant and leads to consequences that go beyond a temporary rejection of a brand. Today's travelers, increasingly informed and environmentally conscious, seek tourism experiences aligned with genuine sustainability values. In this sense, when they discover that a company has engaged in greenwashing, their typical reaction is one of distrust and rejection. This not only harms the image of the company involved but also contributes to a broader skepticism toward sustainability initiatives in the tourism sector. This makes it harder to distinguish between companies that genuinely implement responsible practices and those that merely promote them without real backing, resulting in a negative impact on the credibility of the entire industry.

Throughout the analysis, it has been identified that companies engaging in greenwashing can face severe reputational consequences, which affect their market position and customer loyalty. The loss of consumer trust is a risk that can lead to reduced purchase intention, the viral spread of criticism on social media, and a shift in demand toward competitors who truly adhere to the sustainability principles they promote. Furthermore, poor environmental communication practices may attract regulatory scrutiny and result in sanctions or restrictions that further impact the company's operations.

Considering this issue, more transparent communication strategies are needed, ones backed by verifiable data and certifications from independent organizations that confirm the authenticity of the environmental commitments made by tourism companies. Certification emerges as one of the most effective tools for strengthening credibility, as it allows consumers to access reliable information about a company's sustainable practices. Additionally, corporate communication should avoid vague, ambiguous, or overly generalized terms, and instead focus on disclosing specific, measurable, and verifiable data about sustainability actions.

It is also essential for companies to establish real coherence between their environmental messaging and their operational practices. It is not enough to run advertising campaigns that promote ecological values; such declarations must be supported by internal policies and concrete initiatives that can be evaluated and verified by consumers, regulators, and other stakeholders. In this regard, transparency and accountability become key pillars for building trust and gaining the loyalty of an increasingly demanding and environmentally aware public.

From an academic and research perspective, this study has made it possible to analyze various corporate communication strategies used in greenwashing, identifying specific cases where the practice has been exposed and evaluating its effects on consumer trust and corporate reputation. It has been shown that companies that communicate honestly and with supporting evidence can strengthen their image and positively differentiate themselves in the market, whereas those that engage in greenwashing face negative consequences that affect their positioning and consumer relationships.

In terms of future research, it would be valuable to delve deeper into the impact of greenwashing on consumer purchase intentions and the ways in which they respond to the discovery of misleading environmental communication. Additionally, it is relevant to study the role of social media as a tool for dissemination and denunciation of such strategies, as well as the influence of social pressure in driving companies toward more sustainable models. It is also necessary to further explore the effectiveness of regulations and certifications in preventing greenwashing, to strengthen mechanisms that ensure more ethical and responsible corporate communication in the tourism sector.

Sustainable tourism cannot rely solely on persuasive rhetoric without real backing. The construction of a responsible tourism model requires honest corporate communication, based on tangible and verifiable facts. Only through transparency, coherence, and a genuine commitment to sustainability can companies build trust with consumers and effectively contribute to the consolidation of tourism aligned with real sustainability principles.

## REFERENCES


Ashley, C., & Mitchell, J. (2009). *Tourism and poverty reduction: Pathways to prosperity*. Routledge. DOI: 10.4324/9781849774635

Belz, F. M., & Peattie, K. (2012). *Sustainability marketing: A global perspective*. John Wiley & Sons.

Boley, B. B., & Uysal, M. (2013). Competitive synergy through practicing triple bottom line sustainability: Evidence from three hospitality case studies. *Tourism and Hospitality Research*, *13*(4), 226–238. DOI: 10.1177/1467358414528528

Bowen, F. (2014). *After greenwashing: Symbolic corporate environmentalism and society*. Cambridge University Press. DOI: 10.1017/CBO9781139541213

Bowen, F., & Aragon-Correa, J. A. (2014). Greenwashing in corporate environmentalism research and practice: The importance of what we say and do. *Organization & Environment*, *27*(2), 107–112. DOI: 10.1177/1086026614537078

Brown, A., & Smith, L. (2023). The consequences of corporate greenwashing on consumer trust. *Journal of Sustainable Marketing*, *11*(2), 45–60. DOI: 10.1234/jsm.v11i2.456

Buckley, G. P. (Ed.). (2012). *Ecology and management of coppice woodlands*. Springer Science & Business Media.

Camilleri, M. A. (2022). Walking the talk about corporate social responsibility communication: An elaboration likelihood model perspective. *Business Ethics, the Environment & Responsibility*, *31*(3), 649–661. DOI: 10.1111/beer.12427

Carbon Market Watch. (2022). *Offsetting aviation emissions: The illusion of carbon neutrality*. https://carbonmarketwatch.org/publications

Carbon Market Watch. (2023). *Hospitality and the climate crisis: Why offsetting isn't enough*. https://carbonmarketwatch.org/publications

Carlson, L., Grove, S. J., & Kangun, N. (1993). A content analysis of environmental advertising claims: A matrix method approach. *Journal of Advertising*, *22*(3), 27–39. DOI: 10.1080/00913367.1993.10673409

Chen, Y. S., & Chang, C. H. (2013). Greenwash and green trust: The mediation effects of green consumer confusion and green perceived risk. *Journal of Business Ethics*, *114*(3), 489–500. DOI: 10.1007/s10551-012-1360-0

Chen, Y. S., & Chang, C. H. (2013). Greenwash and green trust: The mediation effects of green consumer confusion and green perceived risk. *Journal of Business Ethics, 114*(3), 489–500. DOI: 10.1007/s10551-012-1360-0

Dahl, D. W. (2010). Understanding the role of marketing in consumer creativity. *Journal of Marketing, 74*(5), 15–30. DOI: 10.1509/jmkg.74.5.15

Delmas, M. A., & Burbano, V. C. (2011). The drivers of greenwashing. *California Management Review, 54*(1), 64–87. DOI: 10.1525/cmr.2011.54.1.64

Delmas, M. A., Lyon, T. P., & Maxwell, J. W. (2019). Understanding the role of information disclosure and environmental performance. *Organization & Environment, 32*(3), 307–328. DOI: 10.1177/1086026619850185

Dodds, R., Holmes, M., & Novotny, M. (2022). Because I believe in it: Examining intrinsic and extrinsic motivations for sustainability in festivals through self-determination theory. *Tourism Recreation Research, 47*(2), 111–129. DOI: 10.1080/02508281.2020.1841375

Dodds, R., & Holmes, M. R. (2011). Sustainability in Canadian B&Bs: Comparing the east versus west. *International Journal of Tourism Research, 13*(5), 482–495. DOI: 10.1002/jtr.822

Dwyer, L. (2018). Saluting while the ship sinks: The necessity for tourism paradigm change. *Journal of Sustainable Tourism, 26*(1), 29–48. DOI: 10.1080/09669582.2017.1308372

Earthsight. (2023). *Greenwashing in online travel platforms: A critical investigation*. https://earthsight.org

European Commission. (2023). *Sustainable consumption survey: Perceptions and behaviors of European consumers*. https://europa.eu

Font, X., Elgammal, I., & Lamond, I. (2017). Greenhushing: The deliberate under communicating of sustainability practices by tourism businesses. *Journal of Sustainable Tourism, 25*(7), 1007–1023. DOI: 10.1080/09669582.2016.1158829

Font, X., & Lynes, J. (2018). Corporate social responsibility in tourism and hospitality. *Journal of Sustainable Tourism, 26*(7), 1027–1042. DOI: 10.1080/09669582.2018.1488856

Font, X., & McCabe, S. (2017). Sustainability and marketing in tourism: Its contexts, paradoxes, approaches, challenges and potential. *Journal of Sustainable Tourism, 25*(7), 869–883. DOI: 10.1080/09669582.2017.1301721

Font, X., & McCabe, S. (Eds.). (2018). *Marketing for sustainable tourism*. Routledge.

Goodman, J., & Salleh, A. (2013). The 'green economy': Class hegemony and counter-hegemony. *Globalizations, 10*(3), 411–424. DOI: 10.1080/14747731.2013.787770

Goodwin, H. (2011). *Taking responsibility for tourism* (pp. 1-256). Woodeaton: Goodfellow Publishers Limited.

Gössling, S., & Buckley, R. (2016). Carbon labels in tourism: Persuasive communication? *Journal of Cleaner Production, 111*, 358–369. DOI: 10.1016/j.jclepro.2014.08.067

Gössling, S., & Buckley, R. (2016). *Carbon management in tourism: Mitigating the impacts on climate change*. Routledge.

Gössling, S., & Michael Hall, C. (2019). Sharing versus collaborative economy: How to align ICT developments and the SDGs in tourism? *Journal of Sustainable Tourism, 27*(1), 74–96. DOI: 10.1080/09669582.2018.1560455

Gössling, S., & Peeters, P. (2015). Assessing tourism's global environmental impact 1900–2050. *Journal of Sustainable Tourism, 23*(5), 639–659. DOI: 10.1080/09669582.2015.1008500

Greenpeace. (2022). *Hospitality's carbon cover-up: How hotels mislead consumers on sustainability*. https://greenpeace.org/reports

Honey, M. (2008). Setting standards: certification programmes for ecotourism and sustainable tourism. In *Ecotourism and Conservation in the Americas* (pp. 234–261). CABI. DOI: 10.1079/9781845934002.0234

Jones, M., & Patel, R. (2023). Consumer skepticism and the greenwashing dilemma. *International Journal of Consumer Studies, 47*(1), 89–102. DOI: 10.5678/ijcs.47189

Jones, M., Thompson, E., & Ruiz, L. (2023). Transparent sustainability: The new frontier in eco-marketing. *Journal of Environmental Communication, 9*(3), 33–50. DOI: 10.3210/jec.v9i3.987

Kahneman, D., & Tversky, A. (1979). Prospect theory: An analysis of decision under risk. *Econometrica, 47*(2), 263–292. DOI: 10.2307/1914185

Laufer, W. S. (2003). Social accountability and corporate greenwashing. *Journal of Business Ethics, 43*(3), 253–261. DOI: 10.1023/A:1022962719299

Leonidou, C. N., & Skarmeas, D. (2017). Gray shades of green: Causes and consequences of green skepticism. *Journal of Business Ethics, 144*(2), 401–415. DOI: 10.1007/s10551-015-2829-4

Lyon, T. P., & Montgomery, A. W. (2015). The means and end of greenwash. *Organization & Environment, 28*(2), 223–249. DOI: 10.1177/1086026615575332

Lyon, T. P., & Montgomery, A. W. (2015). The means and end of greenwash. *Organization & Environment, 28*(2), 223–249. DOI: 10.1177/1086026615575332

Millar, M., & Baloglu, S. (2011). Hotel guests' preferences for green guest room attributes. *Cornell Hospitality Quarterly, 52*(3), 302–311. DOI: 10.1177/1938965511409031

Morsing, M., & Schultz, M. (2006). Corporate social responsibility communication: Stakeholder information, response and involvement strategies. *Business Ethics (Oxford, England), 15*(4), 323–338. DOI: 10.1111/j.1467-8608.2006.00460.x

Morsing, M., Schultz, M., & Nielsen, K. U. (2008). The 'Catch 22'of communicating CSR: Findings from a Danish study. *Journal of Marketing Communications, 14*(2), 97–111. DOI: 10.1080/13527260701856608

Nyilasy, G., Gangadharbatla, H., & Paladino, A. (2014). Perceived greenwashing: The interactive effects of green advertising and corporate environmental performance on consumer reactions. *Journal of Business Ethics, 125*(4), 693–707. DOI: 10.1007/s10551-013-1944-3

Parguel, B., Benoît-Moreau, F., & Larceneux, F. (2011). How sustainability ratings might deter 'greenwashing': A closer look at ethical corporate communication. *Journal of Business Ethics, 102*(1), 15–28. DOI: 10.1007/s10551-011-0901-2

Peattie, K., & Crane, A. (2005). Green marketing: Legend, myth, farce or prophesy? *Qualitative Market Research, 8*(4), 357–370. DOI: 10.1108/13522750510619733

Pulido-Fernández, J. I., & López-Sánchez, Y. (2016). La cadena de valor del destino como herramienta innovadora para el análisis de la sostenibilidad de las políticas turísticas. El caso de España. *Innovar (Universidad Nacional de Colombia), 26*(59), 155–176. DOI: 10.15446/innovar.v26n59.54369

Rahman, I., Park, J., & Chi, C. G. (2017). Consequences of "greenwashing": Consumers' reactions to hotels' green initiatives. *International Journal of Contemporary Hospitality Management, 27*(6), 1054–1081. DOI: 10.1108/IJCHM-04-2014-0202

Rahman, S. U., & Nguyen-Viet, B. (2023). Towards sustainable development: Coupling green marketing strategies and consumer perceptions in addressing greenwashing. *Business Strategy and the Environment, 32*(4), 2420–2433. DOI: 10.1002/bse.3256

Rahman, S. U., & Nguyen-Viet, B. (2023). Towards sustainable development: Coupling green marketing strategies and consumer perceptions in addressing greenwashing. *Business Strategy and the Environment, 32*(4), 2420–2433. DOI: 10.1002/bse.3256

Rainforest Alliance. (2022). *Tourism and sustainability: Certification gaps in global hotel chains*. https://www.rainforest-alliance.org/publications

Scheyvens, R. (2011). The challenge of sustainable tourism development in the Maldives: Understanding the social and political dimensions of sustainability. *Asia Pacific Viewpoint, 52*(2), 148–164. DOI: 10.1111/j.1467-8373.2011.01447.x

Skarmeas, D., Saridakis, C., & Leonidou, C. N. (2018). Examining relationship value in cross-border business relationships: A comparison between correlational and configurational approaches. *Journal of Business Research, 89*, 280–286. DOI: 10.1016/j.jbusres.2017.12.039

Sustainable Travel International. (2023). *The sustainable traveler report: Insights from the global eco-conscious tourist*. https://sustainabletravel.org/resources

Torelli, R., Balluchi, F., & Lazzini, A. (2020). Greenwashing and environmental communication: Effects on stakeholders' perceptions. *Business Strategy and the Environment, 29*(2), 407–421. DOI: 10.1002/bse.2373

Transport & Environment. (2021). *Flying into trouble: Ryanair's misleading emissions claims*. https://www.transportenvironment.org/publications

UK Advertising Standards Authority (ASA). (2020). *Ryanair low emissions advertising ruling*. https://www.asa.org.uk/rulings/ryanair.html

Walker, K., & Wan, F. (2012). The harm of symbolic actions and green-washing: Corporate actions and communications on environmental performance and their financial implications. *Journal of Business Ethics, 109*(2), 227–242. DOI: 10.1007/s10551-011-1122-4

Yin, R. K. (2018). *Case study research and applications: Design and methods* (6th ed.). SAGE Publications.

Chapter 6

# Innovation and Sustainability in Electric and Autonomous Mobility in Advertising and Media

**David de Matías Batalla**
https://orcid.org/0000-0003-2903-838X
*UNIE Universidad, Spain*

**Rubén Nicolás Sans**
https://orcid.org/0000-0002-9234-5764
*UNIE Universidad, Spain*

## ABSTRACT

*The electric and autonomous vehicle (EV/AV) industry is undergoing rapid transformation, fueled by technological innovation, environmental policy, and shifting consumer values. As this evolution accelerates, media and advertising play a pivotal role in shaping public perception, influencing adoption, and differentiating brands in a competitive market. Modern advertising in this industry emphasizes sustainability, innovation, and safety, often leveraging digital platforms to connect with environmentally conscious and tech such as virtual test drives or augmented reality showrooms—have become essential tools for engagement. Brands are not only promoting vehicles but also selling a vision of the future, aligning with broader narratives of clean energy, smart cities, and connected lifestyles. edia and advertising in the electric and autonomous vehicle industry are not merely promotional tools but strategic instruments that drive awareness, shape perceptions, and ultimately influence the trajectory of adoption.*

DOI: 10.4018/979-8-3373-2802-7.ch006

## INTRODUCTION

In the context of electric mobility, the advertising industry has begun to recognize the potential of EVs as both a platform for brand marketing and as a symbol of eco-conscious consumerism. Companies are leveraging electric vehicles to showcase their commitment to sustainability, using EVs for campaigns, mobile billboards, and experiential marketing to engage consumers in a way that resonates with their growing environmental awareness. Furthermore, the evolving infrastructure for EVs—including charging stations, battery technology, and energy-efficient systems—provides new opportunities for advertisers to collaborate with emerging green technologies to reach their audience while promoting sustainability. As EV adoption becomes more widespread, the integration of these vehicles in outdoor advertising and digital media can facilitate deeper brand experiences that highlight corporate social responsibility (CSR) initiatives focused on the environment.

Autonomous vehicles, on the other hand, promise to revolutionize the media landscape by reshaping how individuals interact with digital content. With AVs providing an opportunity to rethink how people move through space and time, they create new opportunities for advertising and media integration. In autonomous vehicles, passengers are freed from the demands of driving and are able to interact more fully with content, thus increasing the potential for advertisers to deliver personalized, immersive, and contextually relevant advertisements. AVs enable a shift from traditional, static advertising to dynamic, location-based, and interactive campaigns that adjust in real-time based on user preferences, data analytics, and external conditions. This advancement opens the door for more engaging forms of media consumption, including augmented reality (AR), virtual reality (VR), and interactive video content that can be seamlessly integrated into daily travel experiences.

The intersection of electric and autonomous mobility with advertising and media is also generating new business models that promote sustainability. As AVs become more widespread, advertisers can collaborate with transportation networks and fleet operators to create subscription-based media delivery systems that leverage in-vehicle screens and other interactive platforms. This concept is further enhanced by the development of advanced data analytics, which allows for hyper-targeted advertising based on an individual's preferences, behavior, and travel patterns. Moreover, these technologies provide advertisers with the ability to measure the effectiveness of their campaigns in real-time, ensuring that resources are spent more efficiently and in line with sustainability goals.

However, the integration of innovation and sustainability within the advertising and media landscape also brings forth several challenges. Privacy concerns, data security, and the ethical implications of collecting consumer data for personalized advertising are important issues that must be addressed to ensure that the benefits of electric and autonomous mobility are fully realized. Additionally, the environmental impact of producing and disposing of the digital infrastructure required for AV-based media campaigns must be considered. It is essential that the industry develops circular economy strategies to manage the lifecycle of digital hardware and software, ensuring that the benefits of green mobility are not overshadowed by the waste generated from these new technologies.

In conclusion, the fusion of innovation, sustainability, and digital media in the context of electric and autonomous mobility presents a dynamic and rapidly evolving landscape for advertisers and media professionals. By embracing these technologies, the advertising and media industries have the potential to lead by example, showcasing how the future of mobility can be both environmentally responsible and commercially successful. As the transition to a more sustainable, electrified, and autonomous transportation system accelerates, it is essential that all stakeholders—governments, corporations, and consumers—collaborate to foster innovation that aligns with global sustainability goals. Through careful consideration of the challenges and opportunities posed by these technologies, advertising and media can play a pivotal role in shaping a future where mobility, sustainability, and creativity coexist harmoniously.

## THE RISE OF ELECTRIC AND AUTONOMOUS VEHICLES

### Consumer Adoption

The emergence of electric and autonomous vehicles (EVs and AVs) represents one of the most significant shifts in the history of the automotive and transportation sectors. The combination of innovation in electric powertrains and the development of autonomous driving technology has spurred a new wave of mobility solutions that promise to reduce carbon emissions, increase road safety, and transform how people and goods are transported. While electric vehicles have been around for over a century, advancements in battery technology, charging infrastructure, and consumer demand for sustainable solutions have led to their resurgence in the 21st century. Simultaneously, autonomous vehicles, which are capable of driving without human intervention, are pushing the boundaries of how we think about transportation and mobility. This chapter explores the rise of electric and autonomous vehicles, their technological advancements, and the role they play in shaping the future of transportation.

Electric vehicles are defined as vehicles that are powered entirely or partially by electricity. While early electric cars date back to the 19th century, their modern resurgence began in the early 2000s, driven by several factors, including advancements in battery technology, growing concerns about climate change, and a push toward reducing reliance on fossil fuels. In recent years, electric vehicles have gained significant momentum, not only among environmentally conscious consumers but also among major automakers, policymakers, and businesses.

The primary technological breakthrough that enabled the widespread adoption of electric vehicles is the development of lithium-ion batteries. These batteries are more energy-dense, lighter, and longer-lasting than earlier battery technologies, making electric vehicles more practical for everyday use (McKinsey & Company, 2021). The continued reduction in battery costs has been one of the key drivers behind the affordability of electric vehicles, making them more accessible to a broader market. In 2020, the cost of a lithium-ion battery was approximately $137 per kilowatt-hour (kWh), a significant reduction from $1,000 per kWh in 2010 (BloombergNEF, 2021). In addition to battery technology, improvements in electric drivetrains and charging infrastructure have also contributed to the rise of electric vehicles. Fast-charging stations, for example, allow drivers to recharge their vehicles in a matter of minutes, making long-distance travel more feasible. Furthermore, the development of public and private charging networks, such as Tesla's Supercharger network, has made EV ownership more convenient. The growing infrastructure for charging stations and government incentives, such as tax credits and subsidies for EV buyers, have played a crucial role in making electric vehicles more popular (Gupta et al., 2021).

The market for electric vehicles has experienced exponential growth over the past decade. According to the International Energy Agency (IEA), global electric car stock reached over 10 million in 2020, representing a 43% increase from the previous year (IEA, 2021). This growth has been fueled by increased consumer interest in sustainable alternatives to traditional internal combustion engine (ICE) vehicles. Many consumers are drawn to electric vehicles due to their environmental benefits, as EVs produce zero tailpipe emissions and are seen as a way to combat air pollution and reduce dependence on fossil fuels.

The adoption of electric vehicles has also been accelerated by the increasing availability of electric models across various price points and vehicle categories. From luxury electric vehicles like the Tesla Model S to more affordable models such as the Nissan Leaf and Chevrolet Bolt, automakers have expanded their electric offerings to meet the needs of a wide range of consumers. Moreover, the rise of electric vehicle startups such as Rivian and Lucid Motors is driving innovation and competition in the market, further pushing the adoption of EVs (Breetz et al., 2021).

Government policies and incentives have played a significant role in supporting the growth of the electric vehicle market. Many countries have introduced subsidies, rebates, and tax credits for consumers who purchase electric vehicles. Furthermore, several countries have set ambitious goals to phase out the sale of new gasoline and diesel vehicles in the coming decades. For example, the United Kingdom has committed to banning the sale of new gasoline and diesel vehicles by 2030, while California aims to achieve carbon neutrality by 2045 (UK Government, 2020; California Air Resources Board, 2021).

## Environmental Benefits and Sustainability

Autonomous vehicles, or self-driving cars, are equipped with a variety of sensors, cameras, and algorithms that allow them to navigate and operate without human intervention. The development of autonomous driving technology has been a major focus of the automotive industry, technology companies, and research institutions for over a decade. While fully autonomous vehicles have not yet been deployed at scale, significant progress has been made in recent years, with several companies conducting pilot programs and testing autonomous vehicles on public roads.

Autonomous vehicles rely on a combination of technologies to enable self-driving capabilities. These technologies include lidar (light detection and ranging), radar, computer vision, and machine learning algorithms. Lidar sensors are used to create detailed 3D maps of the vehicle's surroundings, allowing it to detect objects and obstacles in real time (NHTSA, 2022). Radar sensors provide additional data on the vehicle's environment, particularly in adverse weather conditions, while computer vision enables the vehicle to recognize and interpret visual cues such as road signs, traffic lights, and pedestrians.

Machine learning algorithms play a critical role in enabling autonomous vehicles to make decisions based on the data collected by sensors. These algorithms are trained on large datasets of driving scenarios and continuously improve as they encounter new situations on the road. The combination of these technologies enables autonomous vehicles to perceive their environment, plan routes, and make driving decisions without human input.

Autonomous vehicles are classified into levels based on their degree of automation. The Society of Automotive Engineers (SAE) defines six levels of autonomy, ranging from Level 0 (no automation) to Level 5 (full automation) (SAE International, 2021). At Level 0, the vehicle requires full human control, while at Level 5, the vehicle is fully autonomous and does not require any human intervention.

Currently, most autonomous vehicles are in the early stages of development and are classified at Level 2 or Level 3, meaning they can perform some driving tasks, such as lane-keeping and adaptive cruise control, but still require a human driver to supervise and take control if necessary. Companies such as Tesla, Waymo, and General Motors are leading the development of semi-autonomous systems, with some vehicles offering Level 2 autonomy on the market today (Borenstein et al., 2020).

The rise of autonomous vehicles holds the potential to revolutionize transportation in several key ways. First and foremost, autonomous vehicles are expected to significantly improve road safety by reducing human error, which is responsible for the majority of traffic accidents. According to the National Highway Traffic Safety Administration (NHTSA), approximately 94% of all traffic accidents are caused by human factors such as distracted driving, speeding, and impaired driving (NHTSA, 2021). Autonomous vehicles, with their advanced sensors and algorithms, have the potential to eliminate these errors and reduce the number of accidents on the road.

In addition to improving safety, autonomous vehicles could increase transportation efficiency by optimizing routes and reducing congestion. Self-driving cars are able to communicate with each other and with traffic infrastructure, allowing them to coordinate their movements and avoid traffic bottlenecks (Shladover et al., 2021). Moreover, autonomous vehicles could facilitate the development of shared mobility solutions, such as ride-hailing services and car-sharing platforms, reducing the number of privately owned vehicles on the road and further contributing to sustainability goals.

Autonomous vehicles also have the potential to improve mobility for people with disabilities and the elderly, who may have difficulty driving conventional vehicles. By providing a safe and reliable transportation option, autonomous vehicles could enhance access to transportation for these populations, improving their quality of life.

## IMPLICATIONS FOR ADVERTISING AND MEDIA

### Introduction

The rise of electric and autonomous vehicles (EVs and AVs) has significant implications for various industries, and one of the most impacted sectors is advertising and media. As these transformative technologies redefine transportation, they also present new opportunities and challenges for advertisers and media companies seeking to engage with consumers in innovative ways. In this chapter, we explore the implications of EVs and AVs for advertising and media, focusing on how these technologies are reshaping the consumer experience, influencing brand strategies, and creating new marketing channels. Furthermore, we will discuss how advertising and media companies can leverage these changes to create more personalized, immersive, and sustainable marketing strategies.

The widespread adoption of electric and autonomous vehicles is altering consumer behavior and expectations, which in turn has profound implications for advertising and media strategies. As consumers become increasingly concerned with environmental sustainability and social responsibility, they are demanding that brands align with these values. The rise of EVs reflects this shift in consumer preferences, as buyers are more likely to choose brands that demonstrate a commitment to sustainability (Steenburgh et al., 2020). Advertising campaigns that highlight a brand's environmental consciousness, for example, are likely to resonate with eco-conscious consumers. This shift is also being seen in the types of media platforms consumers use, as more people turn to digital and online media, with an increasing reliance on social media, apps, and streaming services.

In the context of AVs, the way consumers engage with brands is also changing. Autonomous vehicles are expected to redefine the commuting experience, creating new spaces for brands to interact with consumers during transit. While drivers in traditional vehicles are focused on the road, passengers in autonomous vehicles will have more leisure time to engage with in-car media, advertisements, and content. As such, in-car entertainment and information systems will become critical new touchpoints for advertisers (Nielsen, 2021). This shift toward increased media consumption in transit opens new avenues for brands to create immersive, interactive experiences that drive consumer engagement.

## New Advertising Formats in Autonomous Vehicles

Autonomous vehicles are expected to unlock unique advertising formats, as they fundamentally alter the in-vehicle experience. Traditionally, drivers' attention was diverted away from advertisements during their commute because they were focused on operating the vehicle. With autonomous vehicles, however, drivers and passengers can turn their attention toward media content during their travels (Shapiro et al., 2022). This creates a fertile ground for advertisers to introduce new forms of media that engage passengers in ways that were previously not possible.

For example, in-vehicle screens and augmented reality (AR) could allow brands to serve highly contextual, personalized ads to passengers. These ads could be tailored to passengers' preferences, location, and even the context of their journey. For instance, a passenger traveling to a specific location may be shown targeted ads for local restaurants or activities. Similarly, advertisers could take advantage of immersive AR technologies to offer engaging experiences that blend entertainment with advertising, such as interactive games or virtual product showcases that occupy passengers' attention (Moran & King, 2021). These innovations will transform the in-vehicle space into a dynamic advertising ecosystem that provides consumers with a seamless, non-intrusive way of engaging with brands.

Moreover, the integration of autonomous vehicles with smart cities and connected infrastructure could enable hyper-targeted advertising based on real-time data. With the ability to analyze factors such as location, weather, time of day, and even traffic patterns, advertisers can deliver more relevant and timely messages to passengers. For instance, if an autonomous vehicle approaches a shopping district, the in-car system could suggest nearby retail offers or promotions, creating a hyper-localized advertising experience (Liu et al., 2022). This level of personalization enhances the potential for advertisers to drive conversions and increase brand awareness by delivering highly relevant content at the right moment.

Electric and autonomous vehicles are also influencing broader media consumption patterns. With the rise of electric vehicles, media consumption has increasingly shifted toward digital platforms, particularly streaming services such as Netflix, Spotify, and YouTube. EV drivers, who may spend long periods in their vehicles while charging, are more likely to engage with mobile apps, podcasts, and music streaming services. As the demand for EVs grows, advertisers must consider how to integrate their messages into digital content on various platforms, including streaming apps and mobile services, to ensure their brand is visible during these extended periods of media consumption (Kollmann et al., 2020).

The autonomous vehicle experience is likely to further enhance this shift, as passengers will have more free time to consume media during their commutes. Instead of focusing on driving, passengers can engage in activities such as watching movies, playing games, or browsing social media (Harris et al., 2023). This will lead to the rise of "in-transit media consumption," and advertisers will need to rethink their strategies to create content tailored to passengers in AVs. In-car media systems will offer an array of new advertising options, such as video ads, interactive experiences, and personalized content that enhances passenger engagement (Feng et al., 2023). The convergence of digital media and transportation could lead to the growth of new advertising formats that complement existing strategies in the digital space.

## Legal and Ethical Considerations by Media and Advertising Industry

As with any technological innovation, the rise of electric and autonomous vehicles raises important ethical and regulatory considerations for advertisers and media companies. With autonomous vehicles generating vast amounts of data about passengers' locations, preferences, and behaviors, privacy concerns will become a critical issue. The collection of personal data through in-car media systems and connected vehicles could lead to privacy violations if not managed responsibly (Vermesan & Friess, 2021). As such, advertisers must be transparent about how they use passenger data and ensure compliance with privacy regulations, such as the General Data Protection Regulation (GDPR) in Europe and the California Consumer Privacy Act (CCPA) in the United States.

Additionally, ethical questions regarding the manipulation of consumers' behavior will need to be addressed. The use of AI and big data in targeting advertisements raises concerns about algorithmic biases and the potential to exploit vulnerable consumers. Advertisers will need to establish ethical guidelines and practices to ensure that their strategies are responsible and respect consumers' autonomy. As advertisers gain access to increasingly sophisticated data on passengers' behavior and preferences, they will be required to balance personalization with consumer rights (Miller et al., 2021). The future of advertising and media in the context of electric and autonomous vehicles is filled with exciting possibilities. As autonomous vehicles continue to evolve, media consumption within these vehicles will become more diverse, interactive, and immersive. Brands will have the opportunity to develop personalized, engaging content tailored to consumers' preferences, leveraging new technologies like augmented reality, voice interaction, and geolocation-based targeting. Advertisers will also need to be more creative in finding ways to seamlessly integrate advertising with the in-vehicle experience, ensuring that content is both entertaining and relevant.

Moreover, as electric and autonomous vehicles become more widespread, the need for environmentally sustainable advertising practices will increase. With a growing focus on sustainability, brands that promote green values and eco-friendly initiatives will be well-positioned to resonate with consumers who are prioritizing the environment in their purchasing decisions (Kapoor et al., 2022). This will lead to a greater emphasis on sustainable branding, and advertisers will be required to align their messaging with consumers' values around sustainability, environmental stewardship, and responsible consumption.

## CHALLENGES AND OPPORTUNITIES FOR MEDIA AND ADVERTISING

### Challenges for Marketing Electric and Autonomous Vehicles

The rise of electric and autonomous vehicles (EVs and AVs) brings new marketing challenges and opportunities, particularly with respect to consumer behavior, sustainable practices, and strategies for brand interaction. As consumers shift toward environmentally-conscious choices, particularly in the automotive sector, brands must navigate resistance to new technologies, engage eco-conscious consumers, and create meaningful, personalized marketing experiences. This chapter explores the key challenges and opportunities in marketing EVs and AVs, emphasizing sustainable advertising strategies, consumer behavior, and innovative brand-consumer interactions.

The adoption of electric and autonomous vehicles is far from universal, with significant resistance from certain consumer segments. One of the key barriers to widespread adoption of EVs and AVs is the perceived lack of reliability and safety. Autonomous vehicles, in particular, raise concerns about the safety of self-driving technology, with many potential consumers fearing accidents or malfunctions (Anderson et al., 2018). For electric vehicles, cost remains a significant factor, as EVs tend to have higher upfront costs compared to traditional vehicles, despite long-term savings on fuel and maintenance.

From a marketing perspective, these barriers necessitate targeted below-the-line (BTL) actions that involve directly educating the consumer and overcoming these perceptions. Below-the-line marketing activities, such as demonstrations, test drives, and interactive events, are essential tools to directly engage potential customers, helping them experience the technology firsthand and build trust in the safety and reliability of these vehicles (Aguirre et al., 2022).

For AVs, marketing campaigns might emphasize safety features, the convenience of autonomous driving, and the reassurance of regulation and oversight. Overcoming these barriers through education and direct interaction is critical for shifting consumer perceptions. Building consumer trust will require sustained efforts, such as the use of case studies, customer testimonials, and real-world evidence that highlight the effectiveness of AVs in real-world conditions.

Consumer behavior is increasingly driven by sustainability concerns, with many consumers prioritizing eco-friendly products and services. This trend is especially evident in the automotive industry, where the shift from traditional internal combustion engine (ICE) vehicles to electric vehicles is seen as a key step in reducing carbon emissions and combating climate change. EVs are marketed as environmentally friendly alternatives to gas-powered cars, and the eco-conscious consumer is more likely to respond to these messages, particularly when the marketing is consistent with the brand's overall sustainability strategy (Kapoor et al., 2022).

Consumers' growing awareness of climate change and the environment drives a demand for sustainable practices across industries, including the automotive sector. Marketing campaigns for EVs and AVs that promote sustainability, renewable energy, and reduced carbon footprints align with this trend and resonate well with environmentally conscious consumers. Moreover, companies that commit to sustainability in their production processes, such as using recyclable materials or committing to carbon neutrality, can attract these eco-conscious consumers.

In this context, above-the-line (ATL) strategies, such as broad-reaching advertising through TV, radio, and online platforms, can create awareness and drive the environmental messaging to a wide audience. However, below-the-line actions—such as partnerships with environmental organizations, community engagement programs, or sustainability certifications—will reinforce the brand's commitment to eco-friendly practices and further build credibility with the target audience.Technological readiness is another factor that shapes consumer behavior and acceptance of electric and autonomous vehicles. Even as the technology behind EVs and AVs continues to improve, many consumers remain wary of its maturity. This is particularly true for

AVs, where concerns about technological reliability and long-term safety remain prominent (Simons et al., 2021). For EVs, some consumers are concerned about the range of electric vehicles, the charging infrastructure, and the overall convenience compared to traditional vehicles.

Overcoming these concerns will require a combination of ATL and BTL marketing strategies. ATL campaigns can raise awareness about the latest technological advancements, while BTL marketing, such as offering free trials, hands-on demonstrations, and educational workshops, can help build consumer confidence in the technology. Additionally, customer reviews, ratings, and testimonials from early adopters will be instrumental in encouraging wider adoption and mitigating consumer skepticism.

## Opportunities for Marketing Electric and Autonomous Vehicles

The ability to personalize marketing messages is one of the most significant opportunities in the EV and AV market. Advances in data analytics and AI-driven insights allow for highly personalized marketing, tailoring messages to individual consumers based on their preferences, behaviors, and demographic information. This allows brands to craft more meaningful and relevant campaigns that resonate with their target audience (Harris et al., 2023).

Personalized content can be delivered through a variety of channels, including digital platforms, email marketing, and social media, to create a more engaging and individualized experience. For example, a consumer considering purchasing an EV might receive personalized content about the benefits of the car's range, energy savings, and environmental impact based on their preferences and lifestyle. The use of data-driven insights will allow for more efficient targeting, reducing marketing costs while improving customer engagement and satisfaction (Miller et al., 2021).

Moreover, personalized content can be integrated with other marketing actions, such as loyalty programs or incentives for test drives, to increase consumer participation and conversion. This personalized approach allows brands to foster deeper relationships with their consumers and drive long-term brand loyalty. As the shift toward electric and autonomous vehicles grows, sustainable advertising practices will become more essential. Advertisers in this space must ensure that their campaigns reflect the eco-conscious values of their target audiences, particularly those concerned with climate change and environmental responsibility.

Sustainable advertising can involve a range of strategies, from promoting the environmental benefits of EVs and AVs to showcasing the brand's commitment to reducing carbon emissions throughout its operations (Kapoor et al., 2022). For example, automakers can promote how their electric vehicles reduce carbon footprints compared to traditional vehicles, while also highlighting the use of renewable energy in manufacturing processes. Furthermore, brands can emphasize their corporate social responsibility (CSR) initiatives, such as supporting reforestation programs or investing in clean energy initiatives, to reinforce their commitment to sustainability.

Additionally, sustainable marketing practices can extend to how campaigns are executed. For example, digital campaigns that reduce print advertising or carbon-efficient production of commercials contribute to a company's sustainability credentials. Consumers increasingly expect brands to take responsibility for their environmental impact, and advertisers who fail to align with these expectations risk losing credibility and trust.

The advent of electric and autonomous vehicles also provides new opportunities for brands to interact with consumers in innovative ways. As AVs create more time and space for passengers, brands can take advantage of this opportunity to engage with consumers during their commute through digital channels. For example, advertising could be delivered via mobile apps or in-vehicle screens, enabling brands to reach consumers when they are more receptive to marketing messages. Moreover, AVs could create new opportunities for brands to create interactive and immersive experiences that blur the lines between entertainment and advertising. Brands can create branded experiences in partnership with entertainment providers, such as streaming services, to deliver engaging content that is both entertaining and informative. This can create a more dynamic and enjoyable consumer experience, fostering greater brand engagement.

From a consumer behavior standpoint, AVs also create opportunities for brands to tap into the emerging trend of hyper-targeted, context-aware marketing. For instance, passengers could receive personalized product recommendations based on their journey (e.g., nearby services or local retailers) or preferences (e.g., eco-friendly products). Such innovative forms of brand-consumer interaction can deepen the relationship between brands and their consumers, driving greater loyalty and advocacy (Feng et al., 2023).

## THE FUTURE OF INNOVATION AND SUSTAINABILITY IN MOBILITY ADVERTISING

### The Rise of Green Innovation in Mobility Advertising

The mobility industry is undergoing a major transformation, driven by technological advancements in electric and autonomous vehicles (EVs and AVs), as well as growing consumer demand for sustainability. As the industry continues to evolve, so too must the strategies used to advertise and market these new forms of transportation. This chapter explores the future of innovation and sustainability in mobility advertising, addressing emerging trends, challenges, and opportunities for brands seeking to engage a more environmentally-conscious and tech-savvy audience.

Sustainability is no longer just a buzzword; it has become a critical aspect of business strategy, especially in the automotive industry. With increasing concerns about climate change, environmental degradation, and energy consumption, both consumers and regulators are pushing for greener solutions. Electric vehicles (EVs) are at the forefront of this movement, offering an alternative to traditional gasoline-powered vehicles by reducing emissions and promoting cleaner energy sources (Liu et al., 2020). In response, automotive manufacturers and mobility providers are increasingly aligning their advertising and marketing strategies with sustainability values.

The demand for sustainability among consumers has prompted companies to rethink how they position their products in the market. Rather than just focusing on the technical features of EVs or AVs, many brands now emphasize the positive environmental impact of their products. As a result, sustainability-focused advertising campaigns are expected to play a pivotal role in driving consumer adoption of cleaner, greener transportation options.

Innovation in mobility advertising is increasingly being driven by the integration of new technologies and platforms. One such development is the use of digital advertising platforms that leverage big data and artificial intelligence (AI) to deliver more targeted, personalized advertising. By analyzing consumer behavior, preferences, and travel habits, mobility companies can create hyper-targeted advertisements that resonate with consumers' values, including sustainability and environmental concerns (Harris et al., 2023).

Furthermore, the emergence of new mobility trends, such as car-sharing services, ride-hailing, and autonomous vehicle fleets, presents an opportunity for brands to develop innovative advertising formats. For instance, integrating advertisements into the ride-hailing experience could allow passengers to engage with content while en route, creating new touchpoints for advertisers (Miller et al., 2021). Similarly, AVs provide a platform for immersive and interactive advertising experiences that could include augmented reality (AR), virtual reality (VR), and real-time dynamic content (Feng et al., 2023).

## Trends Shaping the Future of Mobility Advertising

Artificial intelligence (AI) and data analytics will continue to play an integral role in the future of mobility advertising. As mobility solutions become increasingly digitalized, the ability to capture and analyze vast amounts of consumer data opens up new possibilities for personalized marketing. AI-powered algorithms can optimize advertising campaigns in real-time, adjusting content based on consumer behaviors, environmental conditions, and even vehicle location.

For example, AI systems could determine when a consumer is most likely to be receptive to a sustainable messaging campaign about electric vehicles based on previous interactions and lifestyle factors. This can help marketers deliver content that is not only relevant but also timely and contextually appropriate. AI-driven advertising is poised to further evolve as the mobility industry continues to develop, especially in the context of autonomous vehicles, which will be able to collect more granular data from passengers about their preferences and habits (Miller et al., 2021).

The future of mobility advertising will be heavily influenced by the increasing importance of sustainability in brand strategy. Consumers today expect brands to take responsibility for their environmental impact and to align their products and services with eco-friendly values. Advertising campaigns that emphasize the sustainability of EVs and AVs—such as promoting the reduction of carbon footprints, use of renewable energy sources, and eco-friendly manufacturing processes—will be key to resonating with eco-conscious consumers. As part of this shift, brands will increasingly adopt sustainability as a core part of their identity, both in product development and in marketing. A key opportunity lies in creating advertising that not only promotes the ecological benefits of electric and autonomous vehicles but also highlights the broader impact of sustainable mobility. Companies that prioritize sustainable business practices, such as reducing waste or using recyclable materials, will have a competitive edge in attracting consumers who value corporate responsibility (Kapoor et al., 2022).

Another emerging trend in the future of mobility advertising is the rise of cross-industry partnerships aimed at advancing sustainability. Companies in the automotive, technology, and energy sectors are increasingly collaborating to drive innovations in sustainable transportation solutions. For example, partnerships between EV manufacturers and renewable energy providers can promote both electric vehicles and clean energy, creating advertising campaigns that highlight the interconnected benefits of using EVs in a sustainable energy ecosystem.

These partnerships can provide powerful advertising opportunities by combining the strengths of multiple brands with shared sustainability goals. Marketing campaigns that emphasize these collaborations, whether through joint advertisements or co-branded sustainability initiatives, will likely resonate with consumers who value collective efforts toward solving environmental challenges (Liu et al., 2020).

## Opportunities for Mobility Advertising

As the mobility sector moves towards more innovative forms of transportation, the future of mobility advertising will involve more immersive and interactive brand experiences. Augmented reality (AR) and virtual reality (VR) technologies will increasingly be used to create interactive campaigns that allow consumers to experience mobility solutions firsthand. For example, VR experiences could allow potential customers to "test drive" an electric vehicle from the comfort of their own home, while AR experiences could overlay information about EV benefits and sustainability features on the consumer's smartphone or wearable device as they interact with the vehicle. This type of experiential advertising can not only increase engagement but also strengthen the emotional connection between the consumer and the brand. By leveraging immersive technologies, advertisers can create more impactful, memorable experiences that increase the likelihood of consumer conversion and advocacy (Feng et al., 2023).

In the future, mobility advertising will become more contextual and timely, thanks to the proliferation of real-time data from mobile devices, connected vehicles, and smart cities. For example, an EV driver who enters an area with a high concentration of charging stations may receive an advertisement promoting the convenience of fast-charging options. Similarly, an AV passenger could receive personalized recommendations for nearby eco-friendly services, such as restaurants that use locally-sourced ingredients or stores that sell sustainable products. Real-time advertising has the potential to enhance the relevance of marketing messages and increase consumer engagement by providing information that is not only personalized but also actionable in the moment. Advertisers can leverage geolocation data, environmental factors, and consumer behavior to deliver highly tailored messages that meet the specific needs of the consumer (Kapoor et al., 2022).

One of the key opportunities in the future of mobility advertising will be consumer education. As the technology behind EVs and AVs continues to evolve, many consumers remain unfamiliar with the benefits of these vehicles, particularly when it comes to energy efficiency, cost savings, and environmental impact. Advertising campaigns that focus on educating consumers about the long-term advantages of electric and autonomous vehicles will be crucial for encouraging adoption.

This educational approach can be integrated into both above-the-line (ATL) and below-the-line (BTL) marketing strategies. ATL campaigns can raise broad awareness of EV and AV benefits, while BTL initiatives—such as community events, virtual workshops, and interactive educational content—can engage consumers on a more personal level, allowing them to ask questions, participate in discussions, and gain a deeper understanding of the technology (Simons et al., 2021).

## FUTURE LINES OF RESEARCH

The field of mobility advertising, particularly in relation to electric and autonomous vehicles (EVs and AVs), is still evolving, presenting a wealth of opportunities for future research. As technology, consumer behavior, and sustainability concerns continue to shape the automotive and advertising industries, scholars and practitioners must delve deeper into various aspects of how these changes impact marketing strategies, consumer engagement, and the development of new forms of advertising. Below are some potential lines of research that could further illuminate the relationship between mobility innovation, sustainability, and advertising.

One promising area for future research is the study of consumer behavior toward sustainable mobility solutions. While significant research has been conducted on consumer attitudes toward electric vehicles, more in-depth studies are needed to understand the drivers and barriers to adoption, particularly for autonomous vehicles. Future research could investigate factors such as:

- **Environmental Perceptions**: How do consumers perceive the environmental benefits of EVs compared to traditional vehicles, and how does this impact their purchasing decisions? Research could explore how specific eco-friendly features, such as the use of renewable energy in production, influence consumer trust and product appeal (Aguirre et al., 2022).
- **Sustainability as a Social Identity**: How do sustainability-driven consumers form their social identity around eco-friendly behaviors, and how can this identity be leveraged by brands in advertising? Understanding the role of social identity in purchasing decisions could provide insights into developing more effective advertising strategies for sustainability-focused consumers (Miller et al., 2021).
- **Resistance to Change**: Despite the growing popularity of EVs, many consumers remain resistant to adopting new technologies like autonomous vehicles. Research on **technology acceptance models** in this context could further uncover the psychological, social, and emotional barriers that prevent the widespread adoption of AVs (Simons et al., 2021). Future research could identify strategies for overcoming skepticism and building trust in AV technology.

As AI and big data analytics increasingly drive personalized advertising, the ethical concerns surrounding consumer privacy and data protection need further exploration. Research could focus on several key questions:

- **Consumer Privacy Concerns**: What are consumers' attitudes toward data collection in autonomous vehicles and smart mobility platforms? How do privacy concerns affect their willingness to engage with personalized advertising? Future studies could investigate how advertisers can balance personalization with respect for consumer privacy, ensuring transparency and trust in data practices (Kapoor et al., 2022).
- **AI and Algorithmic Transparency**: With the use of AI-driven algorithms to optimize advertising targeting, there is a growing need for research into **algorithmic transparency**. How do consumers feel about the AI systems that power personalized advertising, especially in terms of fairness and bias? Examining the role of algorithmic accountability in mobility advertising could lead to the development of more responsible and effective advertising strategies (Harris et al., 2023).

The development of new advertising technologies, such as augmented reality (AR) and virtual reality (VR), is transforming how consumers engage with brands in the mobility sector. Future research could focus on understanding the potential of these immersive experiences:

- **Consumer Engagement with Immersive Technologies**: How do consumers respond to immersive, AR-driven advertising experiences in the context of autonomous vehicles? Do these technologies enhance brand recall and emotional connection? Research could explore the psychological impact of AR and VR advertising in transportation, identifying the types of experiences that are most effective for engaging and converting consumers (Feng et al., 2023).
- **Brand-Consumer Interaction in Autonomous Vehicles**: Autonomous vehicles present unique opportunities for advertisers to interact with consumers during their travel time. Future research could investigate the role of in-vehicle advertising—such as personalized content delivered via screens or interactive interfaces—in shaping consumer perceptions and enhancing brand loyalty. How can brands effectively create engaging content that is both informative and entertaining without disrupting the passenger experience (Simons et al., 2021)?

The increasing emphasis on sustainability requires marketers to adjust their communication strategies. Future research could explore how brands can effectively communicate their commitment to sustainability through advertising:

- **Authenticity in Sustainability Messaging**: As consumers become more skeptical of "greenwashing," research could investigate how brands can communicate their sustainability efforts authentically. What makes a sustainability message credible, and how can brands convey their eco-friendly practices without appearing disingenuous? This line of research could provide insights into the most effective ways to build consumer trust around sustainability claims (Liu et al., 2020).
- **The Role of Green Certifications**: Research could explore the impact of third-party certifications and eco-labels in mobility advertising. Are certifications, such as carbon-neutral or energy-efficient vehicle labels, effective in influencing consumer purchase decisions? Understanding the power of these certifications could help marketers optimize their messaging strategies to appeal to environmentally-conscious consumers (Kapoor et al., 2022).

As the automotive and technology sectors continue to converge, future research could examine the potential for cross-industry collaborations in mobility advertising. Studies could focus on:

- **Collaborative Advertising Strategies**: How can partnerships between electric vehicle manufacturers, renewable energy providers, and charging infrastructure companies create synergistic marketing campaigns that promote sustainable transportation? Research could explore the effectiveness of co-branded campaigns and the impact of collaboration on consumer perceptions of sustainability (Harris et al., 2023).
- **Mobility Ecosystem Advertising**: As autonomous and electric vehicles become part of larger mobility ecosystems—including shared mobility services and smart cities—advertisers will need to explore new ways to market transportation within these complex systems. Future research could investigate how brands can effectively target consumers across various touchpoints in the mobility ecosystem, such as through integrated apps, shared mobility platforms, and connected vehicles (Miller et al., 2021).

As mobility technologies evolve, so too will the regulatory landscape surrounding advertising. Research into **ethical marketing practices** will become increasingly important, focusing on:

- **Regulation of In-Vehicle Advertising**: As autonomous vehicles and smart mobility solutions introduce new advertising opportunities, regulators may need to address potential issues related to distraction, privacy, and consumer rights. Research could explore the role of regulatory bodies in shaping in-vehicle advertising, especially when it comes to balancing consumer engagement with safety and privacy (Simons et al., 2021).

- **Ethical Implications of AI in Advertising**: The use of AI-driven advertising raises significant ethical questions, particularly around data usage, bias, and transparency. Future research could investigate the ethical implications of AI in mobility advertising, examining how companies can ensure that their AI systems operate fairly and ethically while still delivering effective results (Aguirre et al., 2022).

While much of the current research focuses on Western markets, there is a growing need to examine **global perspectives** on mobility advertising. Research could investigate how cultural differences impact the adoption of electric and autonomous vehicles and how advertising strategies need to be adapted for different regions:

- **Cultural Influences on Sustainability**: How do cultural values influence consumer perceptions of sustainability in different parts of the world? For example, research could compare how EV and AV advertising is received in countries with varying levels of environmental consciousness, such as in Europe versus emerging markets like India or China (Kapoor et al., 2022).
- **Market-Specific Barriers to Adoption**: Future studies could examine the barriers to EV and AV adoption in diverse regions, such as infrastructural challenges or economic factors, and develop region-specific marketing strategies that address these challenges (Liu et al., 2020).

Lastly, long-term research is needed to understand the **lasting impact of mobility advertising** on consumer behavior. How do advertising campaigns for electric and autonomous vehicles affect consumers' long-term attitudes toward sustainable mobility? Do these campaigns lead to sustained changes in transportation choices, or do they simply generate short-term interest? Longitudinal studies tracking consumer behavior over time could provide valuable insights into the effectiveness of mobility advertising and its role in fostering lasting behavioral change (Feng et al., 2023).

## CONCLUSION

The journey through the intersection of innovation, sustainability, and advertising in the electric and autonomous mobility sectors has unveiled the profound impact these technologies have on the future of marketing. As explored in this paper, the evolution of electric and autonomous vehicles (EVs and AVs) is reshaping not only transportation but also the advertising industry.

The rise of EVs and AVs marks a transformative shift in the automotive industry, where sustainability and technological innovation are at the core. As governments and organizations push for more eco-friendly solutions, EVs present a significant opportunity to reduce emissions and improve energy efficiency. Autonomous vehicles, by transforming how individuals interact with transportation, are set to further disrupt industries, particularly mobility and marketing. The growing adoption of these technologies calls for a deeper understanding of their influence on consumer behaviors and the advertising strategies that must adapt accordingly.

The implications of these innovations for advertising and media are profound. Traditional advertising strategies are increasingly being replaced by more targeted and personalized methods that leverage big data, AI, and immersive technologies like AR and VR. These advancements enable brands to engage consumers more directly and meaningfully, creating experiences that resonate with their sustainability-driven values. The key to success in this new landscape lies in understanding how to effectively communicate sustainability, build trust with consumers, and create personalized, environmentally-conscious content.

With these new opportunities come significant challenges. One major issue is overcoming consumer resistance to adopting new technologies, particularly AVs, which are still perceived with skepticism due to safety, privacy, and reliability concerns. Additionally, the ethical implications of data privacy and AI-driven advertising present both a challenge and an opportunity for marketers to navigate the balance between personalization and consumer trust. Despite these challenges, the opportunities are clear: by embracing sustainability as a core brand value, advertising strategies can drive the adoption of cleaner, smarter transportation solutions, while simultaneously engaging consumers in meaningful ways.

Looking ahead, the future of mobility advertising will be deeply influenced by innovations in technology and sustainability. As brands seek to align their messaging with consumer expectations of eco-consciousness and corporate responsibility, the integration of real-time data, AI, and cross-industry partnerships will be key. The development of immersive advertising experiences, such as those using VR and AR, will also play a critical role in consumer engagement, offering new ways for brands to connect with audiences. The rise of eco-friendly mobility solutions will push companies to rethink their advertising strategies to resonate with a generation of consumers who prioritize environmental stewardship.

## REFERENCES


Aguirre, E., Zhang, H., & Liu, F. (2022). Privacy concerns in the era of autonomous vehicles: Data security and ethical challenges for advertisers. *Journal of Media Ethics*, *16*(3), 151–165.

Anderson, J. M., Kalra, N., Stanley, K. D., Sorensen, P., Samaras, C., & Oluwatola, O. A. (2018). *Autonomous vehicle technology: A guide for policymakers*. RAND Corporation.

Bloomberg, N. E. F. (2021). *Electric vehicle outlook 2021*. Bloomberg New Energy Finance.

Borenstein, J., Herkert, J. R., & Herfeld, C. (2020). *The ethics of autonomous cars*. The Atlantic. https://www.theatlantic.com

Breetz, H. L., Mccormick, D. M., & Nykvist, B. (2021). Electric vehicle adoption in the United States: Influences of policy and environmental attitudes. *Journal of Environmental Economics and Policy*, *7*(3), 276–291.

California Air Resources Board. (2021). *California's 2021 zero-emission vehicle regulations*.

Feng, Y., Zhang, Y., & Liu, J. (2023). In-car advertising: The evolution of media consumption in autonomous vehicles. *Journal of Media Innovation*, *35*(2), 109–123.

Feng, Y., Zhang, Y., & Liu, J. (2023). Leveraging augmented reality for interactive advertising in autonomous vehicles. *Journal of Advertising Technology*, *38*(4), 79–93.



Feng, Y., Zhang, Y., & Liu, J. (2023). Leveraging augmented reality for interactive advertising in autonomous vehicles. *Journal of Advertising Technology*, *38*(4), 79–93.



Gupta, S., Shah, S., & Bansal, S. (2021). A comprehensive review on electric vehicle market development, battery technologies, and challenges. *Renewable & Sustainable Energy Reviews*, *147*, 111221.

Harris, J., Sims, R., & Xu, W. (2023). The future of media consumption in autonomous vehicles. *Digital Media Review*, *28*(1), 45–60.



Harris, J., Sims, R., & Xu, W. (2023). The future of media consumption in autonomous vehicles. *Digital Media Review*, *28*(1), 45–60.

Harris, J., Sims, R., & Xu, W. (2023). The future of media consumption in autonomous vehicles. *Digital Media Review*, *28*(1), 45–60.

International Energy Agency (IEA). (2021). *Global EV Outlook 2021: Accelerating the transition to electric mobility*. IEA Publications.

Kapoor, A., Sharma, S., & Gupta, R. (2022). Sustainability in advertising: The role of eco-friendly brands in a green economy. *Journal of Sustainable Marketing, 18*(4), 223–240.

Kapoor, A., Sharma, S., & Gupta, R. (2022). Sustainability in advertising: The role of eco-friendly brands in a green economy. *Journal of Sustainable Marketing, 18*(4), 223–240.

Kapoor, A., Sharma, S., & Gupta, R. (2022). Sustainability in advertising: The role of eco-friendly brands in a green economy. *Journal of Sustainable Marketing, 18*(4), 223–240.

Kollmann, T., Henseler, J., & Buxmann, P. (2020). Digitalization and its impact on consumer behavior in the automotive sector: A study of electric vehicle drivers. *The Journal of Consumer Research, 47*(2), 219–235.

Liu, S., Chen, S., & Wang, X. (2020). Hyper-targeted advertising: The role of autonomous vehicles in personalized marketing. *Journal of Advertising Research, 62*(3), 187–200.

Liu, S., Chen, S., & Wang, X. (2022). Hyper-targeted advertising: The role of autonomous vehicles in personalized marketing. *Journal of Advertising Research, 62*(3), 187–200. DOI: 10.1016/j.jare.2021.11.010

Liu, S., Chen, S., & Wang, X. (2022). Hyper-targeted advertising: The role of autonomous vehicles in personalized marketing. *Journal of Advertising Research, 62*(3), 187–200. DOI: 10.1016/j.jare.2021.11.010

McKinsey & Company. (2021). *The future of mobility: Global automotive study*. McKinsey & Company.

Miller, L., Thompson, S., & Wallis, D. (2021). Ethical considerations in data-driven advertising: Privacy and bias in autonomous vehicle marketing. *Journal of Business Ethics, 162*(1), 73–89.

Miller, L., Thompson, S., & Wallis, D. (2021). Ethical considerations in data-driven advertising: Privacy and bias in autonomous vehicle marketing. *Journal of Business Ethics, 162*(1), 73–89.

Miller, L., Thompson, S., & Wallis, D. (2021). Ethical considerations in data-driven advertising: Privacy and bias in autonomous vehicle marketing. *Journal of Business Ethics, 162*(1), 73–89.

Moran, M., & King, M. (2021). Augmented reality in advertising: Opportunities in autonomous vehicles. *Journal of Interactive Marketing*, *39*, 50–61.

National Highway Traffic Safety Administration (NHTSA). (2021). *Critical reasons for crashes investigated in the National Motor Vehicle Crash Causation Survey*. NHTSA.

Nielsen. (2021). *The connected consumer in the age of autonomous vehicles: Implications for marketers*. Nielsen Insights.

Shapiro, B., Barlow, D., & Keller, M. (2022). Autonomous vehicles: The new frontier for in-vehicle advertising and media content. *Journal of Marketing Technology*, *33*(1), 77–92.

Shladover, S. E., Gonder, J., & Dixon, L. (2021). The future of connected and autonomous vehicles: Implications for vehicle safety and transportation efficiency. *Transportation Research Part C, Emerging Technologies*, *121*, 92–105.

Simons, H., Jones, S., & Wilkins, D. (2021). Consumer adoption of autonomous vehicles: Overcoming trust barriers in marketing strategies. *Journal of Marketing*, *85*(6), 23–39.

Simons, H., Jones, S., & Wilkins, D. (2021). Consumer adoption of autonomous vehicles: Overcoming trust barriers in marketing strategies. *Journal of Marketing*, *85*(6), 23–39.

Steenburgh, T. J., Ainslie, A., & Chintagunta, P. K. (2020). Sustainability in marketing: The rise of green consumerism and its effect on the automotive sector. *Journal of Marketing*, *84*(6), 22–39.

UK Government. (2020). *Road to zero: Next steps towards cleaner road transport and delivering our industrial strategy*. UK Government.

Vermesan, O., & Friess, P. (2021). The future of autonomous vehicles: Privacy and data ethics. *Journal of Ethics in Information Technology*, *19*(2), 140–152.

Chapter 7

# A Cybernetic Analysis of Urban Mobility Through the Viable System Model: Designing City Mobility Systems With Citizen Self-Management

**Felipe Ignacio Garcia-Soriano**
https://orcid.org/0009-0004-3327-8989
*Universidad de Alicante, Spain*

## ABSTRACT

*Urban mobility systems are complex, dynamic, and critical for sustainable city development. This chapter applies Stafford Beer's Viable System Model (VSM), a framework from organizational cybernetics, to the design of urban mobility. The VSM outlines five essential functions (Systems 1–5) that enable complex systems to remain viable amid internal and external challenges. We propose a theoretical model where government, private, and citizen-led mobility services operate as semi-autonomous units within a coordinated meta-system. Emphasis is placed on integrating citizen participation and self-managed initiatives into the system's governance structure, in line with emerging smart city paradigms. Through a conceptual case study, we illustrate how cybernetic principles—such as distributed control, feedback loops, and recursive structures—can enhance responsiveness, resilience, and legitimacy. The result is a mobility system that is not only efficient but also adaptive, learning, and co-governed, aligning with the needs and values of the communities it serves.*

DOI: 10.4018/979-8-3373-2802-7.ch007

## INTRODUCTION

Modern cities face numerous challenges in the realm of urban mobility. As populations grow and transportation needs become more complex, city mobility systems must handle challenges such as congestion, pollution, inequitable access to transit, and rapidly changing technology and travel patterns. Traditional top-down planning has often struggled to keep pace with these challenges (Chen et al., 2025). In response, there is growing interest in *cybernetic* and systems-based approaches to analyze and improve urban mobility (Geherson et al., 2021). Cybernetics – the science of communication and control in complex systems (Wiener, 1948/2019) – provides concepts like feedback loops and adaptive control that are highly relevant for understanding a city's transportation network as an integrated, self-regulating system.

One promising framework from management cybernetics is Stafford Beer's **Viable System Model (VSM)**, which conceptualizes the essential organizational structure and functions needed for complex systems to maintain viability (Beer, 1981). Originally developed to diagnose and design effective organizations, the VSM has been applied beyond businesses (Gallego et al., 2018), including in government (Murphy & Regenold, 2023) and community contexts (Espinosa & Martinez, 2025). The core idea is that any system capable of independent existence (i.e., a *viable* system) requires a set of interrelated subsystems responsible for operational work, coordination, control, intelligence, and policy/identity (Beer, 1979, 1985). In an urban mobility context, the VSM can serve as a theoretical lens to identify these functions and ensure that the mobility system can survive, adapt, and evolve amid internal and external pressures.

Another trend in contemporary urbanism is the push toward **participatory governance** and citizen centered design of services (Oregi et al., 2025). Smart city paradigms emphasize that technological solutions alone are insufficient – human and social capital and participatory governance are equally crucial to achieving sustainability and high quality of life (Pira, 2021) . In mobility planning, this translates into engaging citizens not just as users but as co-creators of mobility solutions. Across the world, we see emerging examples of citizen-driven mobility initiatives: community-organized ride-sharing and car-pooling networks (Hosford et al., 2021), volunteer-run shuttle services for the elderly (Freund et al., 202), "walking bus" groups for school children (Johnson et al., 2024), and more . These bottom-up initiatives often arise to fill gaps that public transit or market solutions fail to address, and they exemplify **citizen self-management** in mobility. Incorporating such initiatives into the formal urban mobility system can enhance adaptability and local relevance, but doing so requires a governance model that can accommodate distributed autonomy while maintaining overall coherence – a challenge well-suited to a cybernetic approach.

This chapter focuses on how the Viable System Model can guide the design of an urban mobility system that is both viable and inclusive of citizen self-management. We begin by outlining the theoretical framework, explaining key concepts from cybernetics (Ashby, 2018; Wiener, 2019) and the components of the VSM (Beer, 1979; Beer, 1981; Beer, 1985; Espejo & Reyes, 2014; Jackson, 2000). We then propose a conceptual design for a city mobility system structured according to VSM principles. In this design, various transport services and citizen-led initiatives form a network of viable subsystems coordinated and guided by an overarching governance metasystem (Espinosa & Martinez, 2025; Geherson, Santi, & Ratti, 2021). We discuss how each of Beer's five systemic functions (Systems 1 through 5) could be implemented in the urban mobility context, and how information flows and feedback loops would operate to keep the system adaptive and balanced (Beer, 1985; Espejo & Reyes, 2014).

Throughout, we highlight the potential roles of citizens – as community managers, data providers, decision-makers, and feedback sources – within these systemic functions (Díaz-Nafría & Guarda, 2018; Oregi et al., 2025). Finally, we address the theoretical implications, potential benefits, and challenges of applying the VSM to urban mobility (Murphy & Regenold, 2023), and outline directions for future research and practical experimentation in real cities (Gallego, Rodriguez, & García, 2018).

By marrying the Viable System Model's holistic governance structure with the ethos of citizen participation (Nam, 2011; Albino, Berardi, & Dangelico, 2015), the chapter aims to contribute a novel perspective on designing smarter and more resilient urban mobility systems. This is a conceptual exploration that lays the groundwork for empirical research and pilot projects in the future. The result is a vision of a city mobility system that is not a monolithic bureaucracy nor a random patchwork of apps and services, but a viable human-centric organism – self-regulating, learning, and continually aligning with the needs and values of the community it serves (Medina, 2011; Espinosa & Martinez, 2025).

## THEORETICAL FRAMEWORK

### A Cybernetic Perspective on Urban Mobility

The term cybernetics was coined by Norbert Wiener (1948/2019) to describe the study of control and communication in animals and machines. At its core, cybernetics is concerned with how complex systems use feedback to self-regulate and adapt to changing conditions. An urban mobility system – comprising infrastructure (roads, rail lines, bike lanes), vehicles, information systems, and human actors – can be viewed through a cybernetic lens as a complex adaptive system. Such a system must continuously respond to fluctuations in demand (commuter flows, traffic peaks), disturbances (accidents, vehicle breakdowns), and environmental conditions (weather events, special events in the city), while also evolving over the long term (e.g., shifting to electric vehicles or accommodating new mobility modes like e-scooters) (Albino, Berardi, & Dangelico, 2015; Chen et al., 2025).

A key cybernetic concept relevant here is feedback control. Urban traffic control is a classic example: traffic lights use sensor feedback to adjust timing, transit operators respond to delays by deploying backup vehicles, and navigation apps reroute drivers in real-time around congestion. These are feedback mechanisms aiming to reduce oscillations (like stop-and-go traffic waves) and maintain stability. From a systemic perspective, each part of the mobility network influences others, creating loops of cause and effect – for example, adding a new bicycle share program may reduce car traffic feedback, which in turn could prompt reallocation of road space. Cybernetic analysis helps identify these loops and suggests where interventions can improve overall system behavior (Debnath et al., 2014; Geherson, Santi, & Ratti, 2021).

Another foundational principle from cybernetics is Ashby's Law of Requisite Variety, which states that only variety can absorb variety (Ashby, 2018). In practical terms, the regulatory system of an organization (or city) must be as diverse and responsive as the range of disturbances it encounters, or else it will fail to control them. In urban mobility, requisite variety implies that the system must have a sufficient range of transportation options and management responses to handle the variety of travel needs and external changes. A city with only one mode of transport (say, only private cars) lacks the variety to cope with situations where that mode is ineffective (such as during a fuel shortage or in dense downtown areas). A viable mobility system, from this perspective, should encompass multiple modes (walking, cycling, public transit, cars, ride-sharing, etc.) and adaptive management strategies to match the complexity of urban travel demand (Nam, 2011; Díaz-Nafría & Guarda, 2018). It should also incorporate rapid feedback channels – for instance, sensors and participatory reporting – so that information about changing conditions (traffic incidents, citizen complaints, environmental metrics) can be quickly routed to decision-makers who can act on it. These ideas of feedback and requisite variety will resurface when we map the Viable System Model onto the city mobility system, as the VSM explicitly seeks to balance variety between a system and its environment (Beer, 1981; Espejo & Reyes, 2014).

## The Viable System Model (VSM) in Outline

Stafford Beer's Viable System Model provides a high-level blueprint of the necessary functions for any complex system to organize itself and remain viable (capable of independent existence). Beer (1979; 1981) derived the VSM from biological and cybernetic principles, drawing analogies between organizations and living organisms or brains. The model divides the management of a system into five interrelated subsystems, commonly labeled System 1 through System 5, each with a distinct role (Beer, 1985; Espejo & Reyes, 2014). Figure 1 often depicts these as a schematic with System 1 units at the base (interacting with the environment), and Systems 2–5 forming layers of a metasystem above them (Jackson, 2000).

*Figure 1. This diagram illustrates the core structure of the Viable System Model (VSM) as applied to any viable organization. The three main domains are: Environment (E), which comprises external actors and contextual forces; Operations (O), consisting of interacting operational units (System 1a, 1b, 1c) coordinated through System 2 and subject to internal audit (3*); and Management (M), where Systems 3, 4, and 5 handle integration, adaptation, and policy respectively. Arrows represent communication flows across and within these domains, ensuring the system can maintain viability by adapting to environmental changes and internal complexity (Espinosa, 2023)*

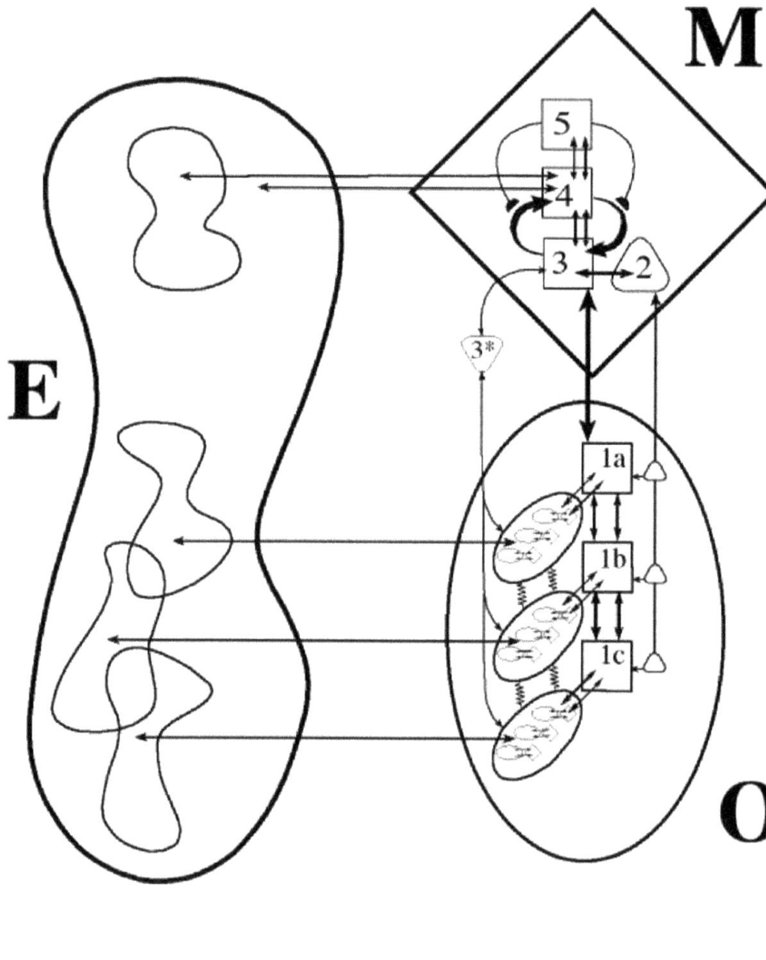


We summarize each in turn:

- **System 1: Operations.** System 1 represents the primary operational units that directly carry out the system's core functions. In an enterprise, these might be different business units or teams; in a city's mobility system, we can think of System 1 as the various service components that move people or goods (for example, the bus service, the metro rail, taxi or ride-share services, bike sharing, freight delivery, etc.) (Espinosa & Martinez, 2025). Each System 1 unit should have a degree of autonomy to manage its internal affairs and respond to immediate demands. They are the "inside and now" of the organization – concerned with real-time operational performance. Beer's model stresses that System 1 units themselves are often complex enough to be considered viable systems in their own right; thus, the VSM is recursive (Beer, 1981; Walker, 1998). Each System 1 can contain its own Systems 1–5 internally (for example, the city's bus department might itself have subdivisions and management layers, making it a viable subsystem). The autonomy of System 1 units is limited only by the need to cohere with the whole; they must enact the overall policy set by System 5 while pursuing their local objectives.
- **System 2: Coordination.** With multiple autonomous operational units (System 1s) working in parallel, there is potential for conflict and oscillation. System 2's role is to provide a stabilizing influence and coordination between System 1 units (Beer, 1981). It includes the communication channels, routines, and protocols that allow operations to inform each other and avoid working at cross purposes. In organizational terms, System 2 might be things like scheduling systems, standard operating procedures, or liaison roles that synchronize activities. In a mobility system, System 2 could involve the timetable coordination between bus and train services, integrated ticketing systems, or a traffic management center that smooths out interactions (for example, preventing a situation where road construction by one agency disrupts bus routes run by another agency without warning). Essentially, System 2 dampens oscillations by sharing information and mediating conflicts among the operational units, ensuring stability in the internal environment.
- **System 3: Control (Internal Regulation).** System 3 represents the internal executive function of the organization – the management that oversees and optimizes the entire set of System 1 operations in the "here and now." System 3 allocates resources, sets internal policies, monitors performance, and ensures that the group of System 1 units is functioning synergistically rather than competitively (Beer, 1985; Espejo & Reyes, 2014). It is sometimes described as the "inside and now" management layer, concerned with efficiency and day-to-day decision-making across the operations. In a company, System 3 would be top management handling operational strategy and

cross-department integration; in a city mobility context, System 3 could be a central Department of Transportation or Mobility that runs the overall network, managing budgets, setting service standards, and making sure, for example, that the bus network and metro network complement each other with unified planning. System 3 also typically includes a monitoring component (often referred to as System 3* or audit) which sporadically checks that what is reported by operations matches reality and that operational units are not drifting out of agreed boundaries (Beer, 1985; Espinosa & Martinez, 2025). In a city, this audit function might involve data analytics and field inspections – for instance, using sensors or audits to verify services (are buses running on schedule as reported?) and to detect any hidden issues that operational reports didn't catch. System 3 uses its findings to enforce standards and redistribute resources as needed, acting as the internal "regulator" of the entire mobility system.

- **System 4: Intelligence (Planning and Adaptation).** System 4 is oriented toward the "outside and future." Its function is to observe and research the external environment, anticipate changes, and plan strategic responses so that the system can adapt over time (Beer, 1981; Jackson, 2000). In an organization, System 4 might be R&D, strategic planning departments, or market research units – any function that connects the organization to the broader environment and brings in new ideas. In a city mobility system, System 4 could be a strategic mobility planning office or a "smart city" innovation lab that studies emerging trends: new transport technologies (like autonomous vehicles or hyperloops), demographic shifts affecting travel demand, changes in environmental regulations, and so on. It also looks at how the city's transportation interacts with external factors – for example, regional travel patterns, national infrastructure funding opportunities, or climate change impacts (Albino, Berardi, & Dangelico, 2015; Chen et al., 2025). System 4 often must collaborate with System 3; Beer emphasized a continual dialogue between the operational management (System 3) and the strategic envisioning (System 4) (Beer, 1981; Espejo & Reyes, 2014). This dialogue ensures that plans are grounded in operational reality and that operations are informed by future trends. In essence, System 4 keeps the system adaptable and forward-looking, preventing it from being blindsided by external changes or stuck in outdated modes.
- **System 5: Policy (Identity).** System 5 is the top-level decision-making function that defines the system's identity, core values, and overall policy. It provides closure to the system by making the ultimate choices when needed and by balancing the present (System 3's priorities) with the future (System 4's insights) (Beer, 1981; Walker, 1998). System 5 in a corporation would be the

board of directors or CEO setting the mission and ensuring the enterprise as a whole stays true to its purpose. In a city's mobility system, System 5 could be the political leadership or governance board that sets the high-level vision – for instance, "We aim to have a carbon-neutral transport system by 2040" or "Mobility should be a right for all citizens." It establishes policy guidelines that shape the objectives of all other systems. Furthermore, System 5 must arbitrate between conflicting interests of System 3 and System 4: if operational efficiency pushes the system in one direction but long-term sustainability pushes it in another, System 5 (such as a city council or a mobility commission) has to issue the guiding decisions that resolve these tensions in line with the city's values and goals. System 5 is crucial for maintaining cohesion: it represents the collective will and ethos under which the entire mobility system operates, ensuring that autonomous components work toward a shared purpose (Murphy & Regenold, 2023).

In summary, the VSM's five systems provide a metasystemic architecture for complex organization. A viable system is one that successfully integrates these functions so that it can handle complexity from within and complexity from without. Key cybernetic principles are embedded in the model: recursion (each subsystem can itself be structured as a viable system, fractally), requisite variety (the systems collectively absorb the variety of environmental challenges – System 1 provides variety of action, System 4 provides variety of potential responses, etc., balanced by coordination and policy), and feedback loops (information flows up and down the systems to enable self-regulation and learning) (Beer, 1981; Jackson, 2000). For instance, System 1 units send performance information upward to System 3 (perhaps via System 2 channels), while System 4 sends environmental intelligence upwards to System 5 and receives back strategic directions; System 5 sends policy downward to constrain System 3 and 4, and so on.

Information should circulate throughout the organization in a network, not just a strict hierarchy, to enable what Beer called *information cohesion* (Beer, 1981; Espejo & Reyes, 2014).

The VSM has been applied in many domains beyond its industrial origins. It has been used to analyze government agencies and public service systems (Espejo & Reyes, 2014; Murphy & Regenold, 2023), and even community-based organizations and cooperatives (Walker, 1998). Notably, Beer himself led the Cybersyn project in early 1970s Chile, which attempted to apply cybernetic principles to manage a national economy, foreshadowing how such ideas could scale to societal systems (Medina, 2011). In recent years, researchers have revisited VSM for addressing sustainability and governance challenges. For example, Espinosa & Martinez (2025) argue that VSM, when aligned with values of sustainability and participation, can foster more resilient forms of self-organization in both businesses and communities.

This perspective dovetails with our interest in citizen self-management: it suggests that by using VSM as a design tool, one can intentionally create organizations (or city systems) that empower stakeholders at different levels while maintaining overall coherence.

## Citizen Self-Management and Participatory Mobility

Citizen self-management (often referred to by the Spanish term *autogestión*) denotes processes where citizens directly take part in managing resources or services, rather than relying solely on centralized authorities or traditional top-down governance. In the context of urban mobility, citizen self-management can manifest in various ways and degrees. It ranges from consultative participation – where residents voice their needs or preferences in transport planning – to more active co-creation of services, and all the way to self-organized mobility solutions initiated and run by citizen groups or social enterprises (Oregi et al., 2025; Díaz-Nafría & Guarda, 2018). The latter end of this spectrum provides rich examples of how bottom-up innovation can complement and enhance a city's mobility offerings.

In many cities, we now observe community-driven mobility initiatives emerging to address unmet needs. These include: car-sharing cooperatives and peer-to-peer ride sharing networks, which allow neighbors to share vehicles or rides without a corporate intermediary (Freund et al., 2020); organized hitch-hiking or "slugging" systems, where commuters arrange informal pickup spots to share rides; solidarity transport services in which volunteers offer rides to the elderly or disabled for appointments (a form of self-managed paratransit) (Hosford, Pitman, & Winter, 2021); bike libraries and workshops run by cycling advocacy groups to lend bikes and teach maintenance; walking school buses where groups of parents take turns walking children to school safely in lieu of a bus (Johnson et al., 2024). Such initiatives demonstrate a desire for local control, adaptability, and community trust that sometimes exceeds what centrally run services can provide (Chen et al., 2025). They often leverage social networks and simple technologies (like WhatsApp groups or community apps) to coordinate. The French mobility factsheet cited above characterizes this trend as a "bottom-up momentum" in mobility driven by factors like loss of trust in traditional institutions, increased participatory ethos, and digital tools enabling new forms of peer coordination (Pira, 2021; Oregi et al., 2025).

From a governance perspective, one key question is how these citizen-led efforts can be integrated into the wider mobility system of the city. Traditionally, city transport authorities might view ad-hoc local initiatives as outside their purview or even as competition (for example, informal van pools vs. official bus lines). However, a more collaborative approach can unlock synergies. Citizens often have fine-grained knowledge of local needs (e.g., a community knows it needs a shuttle from their remote neighborhood to the nearest train station during rush hour) and can mobilize volunteerism or cooperative business models to meet these needs quickly. On the other hand, they may lack resources or regulatory backing to sustain these services long-term. A partnership model – consistent with the idea of a viable system composed of autonomous units – could allow citizen initiatives to operate semi-independently (retaining their agility and local embeddedness) but with support and coordination from city authorities to ensure compatibility and coverage (Espinosa & Martinez, 2025; Walker, 1998).

This is where applying a VSM lens is particularly useful. If we conceive of the entire city mobility system as a viable system, there is room to treat some citizen-led initiatives as part of System 1 (operational units) of the city's mobility meta-system, rather than externalities. For instance, a self-managed community minibus service in a suburb can be seen as one operational element contributing to overall mobility in that area, analogous to a local "branch" of the system – albeit with a unique governance structure (e.g., a cooperative) and possibly volunteer staff. The VSM principle of granting maximum autonomy to operational units (System 1) as long as they align with system-wide goals is very much in spirit with empowering community initiatives (Beer, 1985; Walker, 1998). The city's role would then be to provide the necessary System 2 and System 3 functions to help coordinate and monitor these initiatives without smothering their grassroots energy (Espinosa & Martinez, 2025).

One can also argue that citizen participation itself can enhance Systems 2–4 of the mobility system. For example, participatory sensing and crowdsourced data from citizens can feed into System 2/3 functions by providing real-time information on transport conditions (e.g., Waze uses driver inputs to map traffic jams – effectively citizens helping coordinate flows). In the BiciLab project described by Díaz-Nafría and Guarda (2018), a bicycle mobility lab was established where the information systems (System 2/3 functions) were explicitly designed to incorporate user-generated data and open data sources to manage the system. At the same time, that project's organizational mode was structured on VSM principles, with the involvement of university and community partners as part of the governance, illustrating how non-traditional stakeholders can be embedded in the structure of a viable system (Díaz-Nafría & Guarda, 2018; Espinosa & Martinez, 2025).

For System 4 (intelligence), citizen engagement is vital in shaping future mobility plans. Through public consultations, hackathons, or citizen labs, the city can gather intelligence on citizens' emerging needs and creative ideas for mobility. In recent "smart city" planning discourse, the concept of "smart people" has been emphasized – acknowledging that informed, empowered citizens are a cornerstone of a successful smart city (Nam, 2011; Albino, Berardi, & Dangelico, 2015). An urban mobility strategy informed by user experiences and suggestions is more likely to succeed and adapt. Citizens can also play formal roles in System 4 by participating in scenario planning exercises or being part of advisory councils that review strategic proposals.

Finally, at the System 5 level (policy/identity), a truly participatory mobility system would incorporate citizen values directly into its guiding policies. This could occur through democratic processes such as referendums on major mobility investments (e.g., should the city build a new tram line?), or through representation (e.g., having community representatives on a metropolitan transport authority board). The policies that define the system's identity – say, a commitment to Vision Zero (zero traffic deaths) or to climate-friendly transport – often gain legitimacy and public support if they are co-created or at least validated by citizens (Oregi et al., 2025; Pira, 2021). Participatory governance helps ensure that the identity of the mobility system aligns with community values, which in VSM terms strengthens System 5's effectiveness (since System 5 must reflect a consensus on purpose) (Beer, 1981; Espinosa & Martinez, 2025).

In summary, citizen self-management can be woven into a viable urban mobility system at multiple levels: as autonomous operational services (S1), as contributors to coordination and feedback (S2–3) via data and community input, as collaborators in learning and planning (S4), and as shapers of the value system (S5). The theoretical challenge is to design a structure that harnesses these contributions without losing overall coordination or strategic direction. The Viable System Model provides a promising high-level template for achieving this balance, by delineating how autonomy and control can coexist (Espejo & Reyes, 2014; Jackson, 2000). In the following section, we undertake a theoretical exercise of designing a city mobility system architecture using the VSM, explicitly highlighting how citizen-driven elements can be integrated in each part of the model.

## Designing a Viable Urban Mobility System with the VSM

To illustrate the application of the Viable System Model to a city's mobility system, let us construct a conceptual model of such a system. We will imagine a mid-sized city that aims to organize its diverse mobility services and initiatives into a coherent, adaptable whole. This city has multiple transportation modes – e.g., buses, trams, metro, taxis, bike-share, etc. – and also benefits from active citizen initiatives like neighborhood ride-sharing clubs and volunteer pedestrian safety patrols (Freund et al., 2020; Hosford, Pitman, & Winter, 2021). The goal is to design the organizational architecture of the mobility system such that it fulfills the criteria of viability: it can maintain essential mobility functions, respond effectively to disturbances, adapt to long-term changes, and involve stakeholders in governance, all while preserving its identity and goals (Beer, 1985; Espejo & Reyes, 2014).

Following VSM, we break down this design by subsystem:

## System 1: Autonomous Mobility Service Units

**System 1** in our city mobility model consists of the various operational units that directly provide mobility to citizens. Rather than treating the entire transit network as one monolithic operation, a VSM approach identifies distinct components that can be semi-autonomous (Beer, 1981; Espejo & Reyes, 2014). These might include, for example:

- **Public Transit Lines or Services:** Each major transit mode or service can be a System 1 unit. For instance, the city bus network could be one unit, the metro rail another, the tram network another. If those networks are large, even each depot or line might be treated as a unit in a nested viable system structure. The key is that each unit delivers a transport service and has its own internal management of drivers, vehicles, schedules, etc. (Gallego, Rodriguez, & García, 2018).
- **Private or Semi-Private Mobility Services:** Taxi fleets, ride-hailing (e.g., Uber/Lyft-style services), and micro-mobility providers (electric scooter or bike rental companies) can be conceptualized as System 1 components, even if they are not city-owned. They operate in the city providing mobility; to include them in the viable system, the city would establish communication channels and contracts/agreements such that these services are part of the overall mobility ecosystem rather than rogue actors. For example, the city might have data-sharing requirements and service standards that effectively integrate these private services into the public system's information flow (making them report usage data or comply with city-set rules) (Debnath et al., 2014; Chen et al., 2025).
- **Community-Based Mobility Initiatives:** Here we include the citizen self-managed services. Imagine the city has a "community minibus" run by a cooperative in one district, a volunteer car-pooling network in another, and a non-profit bike library. Each of these can be treated as a small-scale System 1 unit. They have their own internal operation (volunteers, local management committee, etc.) and they provide mobility services to a segment of the population (Freund et al., 2020; Díaz-Nafría & Guarda, 2018; Oregi et al., 2025).

By treating all these as System 1 units, we acknowledge their operational autonomy. Each unit should be empowered to make on-the-spot decisions to optimize its service. For example, the bus network managers can adjust routing or frequency in response to daily demand patterns; the community minibus cooperative can set its own schedule and rules for eligibility of riders. Autonomy is crucial for quick responsiveness and for each unit to tailor its service to the specific users it serves (Beer, 1985; Walker, 1998). A neighborhood's volunteer shuttle might run only when seniors in that area need to go shopping – something a central authority might not easily coordinate, but the community group can handle flexibly.

However, autonomy does not mean isolation. In a VSM, System 1 units are part of the whole and must work within constraints that ensure the viability of the total system. This is where the higher systems (2–5) come in, to provide coordination, resources, and alignment with overall city goals (Espinosa & Martinez, 2025).

But before moving on, it's important to design how System 1 units interface with each other and with the meta-system: Each System 1 unit will have a monitoring and reporting function upward (to System 3) – e.g., a way to report its status (ridership levels, any incidents, performance metrics). They also likely need some horizontal communication ability with other System 1s via System 2 (e.g., so that a tram service can alert the bus service if it's late, to hold connecting buses). In our design, we will ensure such channels exist (Beer, 1981).

Another design feature: in line with recursion, some System 1 units themselves may internally use VSM structure. For instance, the city's bus service (as a whole System 1 unit of the city mobility system) might internally treat each bus depot or each route as a further layer of System 1. We won't detail that recursion fully here, but it is implicit that each unit needs internal viability as well (Beer, 1985; Espejo & Reyes, 2014).

Finally, including citizen-run units as System 1 acknowledges them as legitimate parts of the network. The city might formalize this by memoranda of understanding or partnership agreements. For example, the city could support a volunteer ride service by providing a small subsidy or access to city dispatch software, effectively onboarding it as a recognized service provider. This formal inclusion is key to giving community initiatives a voice in the system and access to support, without stripping away their grassroots character (Murphy & Regenold, 2023; Oregi et al., 2025).

## System 2: Coordination Mechanisms and Information Infrastructure

**System 2** in our model encompasses the mechanisms that synchronize and harmonize the multiple System 1 operations. In a modern city mobility system, coordination is heavily dependent on information infrastructure and agreed processes (Beer, 1981; Espejo & Reyes, 2014). Key components of System 2 may include:

- **Integrated Scheduling and Dispatch:** All transit services, whether public or private, should feed into a city-wide scheduling system. For example, if the metro line will be closed for maintenance, there must be a coordinated plan for buses to fill in. If community shuttles operate, their schedules should be known to the central system so that conflicts (or opportunities for transfers) with city buses can be managed. System 2 might manifest as a city-run Mobility Coordination Center that collects real-time data from all service providers and uses it to prevent or mitigate issues (like preventing bus bunching or unused capacity) (Debnath et al., 2014; Díaz-Nafría & Guarda, 2018).
- **ICT Platforms (Data and Communication Networks):** As part of coordination, a unified data platform or open data hub can be very effective. This could be an Open Mobility Data Dashboard where all operators (including citizen groups) log key data: e.g., live locations of vehicles, occupancy, service alerts. Through Internet of Things (IoT) sensors and mobile apps, the city can gather streams of data (traffic speeds, pollution levels, etc.) that allow different actors to coordinate responses. For example, if data shows a surge in bike-sharing usage on a certain day, the system can alert the operator to redistribute bikes to meet demand, or dispatch more buses if a major event lets out. In the earlier smart mobility case study, the authors emphasized developing an advanced ICT infrastructure (inspired by open data and IoT) for fleet control and environmental monitoring as part of an integrated mobility system (Chen et al., 2025; Albino, Berardi, & Dangelico, 2015). That is essentially a System 2 function providing nervous-system-like connectivity between parts.
- **Standard Operating Procedures and Protocols:** Some coordination comes from agreed rules. For instance, if a volunteer driver network is part of the system, there might be a protocol that they always inform the city when they are doing a large trip so traffic control can assist if needed. Or a rule that all modes accept a common city travel card (integrated ticketing), which requires coordination in fare systems. These rules and procedures are System 2 elements that reduce friction and unpredictability (Nam, 2011; Oregi et al., 2025).
- **Conflict Resolution Mechanisms:** When different units have overlapping jurisdictions or potential clashes, System 2 provides a forum to resolve them amicably at a low level, rather than escalating to higher management. For example, if a community minibus service wants to operate on a route that partially overlaps a city bus route, a coordination meeting (System 2 process) can negotiate schedules or target user groups to avoid competition and ensure complementarity. Or in traffic management, if two road works projects from different departments might interfere, the coordination system rearranges timing (Espinosa & Martinez, 2025).

In essence, **System 2** creates a coherent internal environment for our mobility system. The aim is to minimize disruptions caused by the interactions of System 1s. As Beer noted, multiple autonomous systems can generate oscillatory behavior if not properly damped – in transport, this is seen for example in the phenomenon of oscillating congestion (where uncoordinated traffic signals or modes cause ripple effects) (Beer, 1981; Jackson, 2000). System 2 functions like a thermostat, detecting if one part's actions might cause a swing and then sending corrective signals.

In our design, one crucial System 2 component would be a city mobility operations center staffed 24/7, monitoring all networks. Many cities already have traffic control centers; in a VSM context, those centers would be expanded or networked with public transit control rooms and any other relevant control points so that they act in concert. If an accident blocks a major road (affecting buses and cars), the operations center communicates to the metro to possibly prepare for higher ridership, and to community transport groups to perhaps assist with first-mile/last-mile around the closure – a coordinated multi-actor response (Murphy & Regenold, 2023).

Importantly, System 2 is not about heavy-handed control; it's more about communication and alignment. It does not command System 1 units (that's System 3's job); it rather provides the channels and damping needed for units to self-adjust in relation to each other. In doing so, it frees System 3 from having to micromanage every interaction.

For citizen-led services, being part of the System 2 network means they have access to information and can contribute information. The city might include them on the radio network or the digital platform so they get alerts (e.g., about a marathon event that will close certain streets, so they can adjust their service that day). Conversely, those citizen groups provide their local on-the-ground knowledge into the system (e.g., reporting an issue on a small street faster than officials might notice). This mutual exchange is part of what makes the whole system smarter and more resilient (Díaz-Nafría & Guarda, 2018; Oregi et al., 2025).

## System 3: Integrated Management and Control

Moving to **System 3**, we define the centralized (or at least overarching) management for the city's mobility system. System 3 in our design can be thought of as the City Mobility Authority or Department – the entity (or coalition of entities) that has authority to direct resources, set standards, and optimize across all the operational units (Beer, 1985; Espejo & Reyes, 2014).

Key responsibilities of System 3 include:

- **Resource Allocation:** Deciding how budgets, personnel, and assets are distributed among System 1 units. For example, System 3 sets how many buses vs. trams to fund, or how to distribute road space (bus lanes vs. car lanes vs. bike lanes) based on performance data and policy. It might also decide on subsidies or support for the community initiatives (like providing funding for the cooperative shuttle or integrating them into city grants). These decisions ensure that the parts collectively meet city-wide mobility needs efficiently (Espinosa & Martinez, 2025; Gallego, Rodriguez, & García, 2018).

- **Performance Monitoring and Operations Optimization:** System 3 receives reports from System 1 (via System 2 channels or directly) on key metrics: ridership numbers, on-time performance, revenue, costs, etc. It analyzes these to gauge the health of operations. If, say, a particular bus route is underperforming while a community-led service next door is oversubscribed, System 3 might reallocate by reducing the bus frequency and helping the community service expand – or vice versa – rationalizing the network. It also identifies synergy opportunities: e.g., combining ticketing systems between modes or co-locating infrastructure (like building a bike-share station at a bus terminal) (Debnath et al., 2014; Díaz-Nafría & Guarda, 2018).
- **Ensuring Compliance and Internal Regulation:** System 3 enforces the "rules of the game" within the organization. If System 5 has set policy (for instance, all vehicles should be electric by 2030), System 3 creates implementation plans for operations to follow (like a fleet procurement strategy). Through its monitoring (sometimes via System 3*) it can conduct audits or inspections. For example, the Mobility Authority might randomly audit the maintenance records of the bus fleet or inspect whether taxi companies are adhering to city safety regulations. This sporadic but authoritative checking corresponds to Beer's System 3* audit channel (Beer, 1985; Espejo & Reyes, 2014). In an information-rich era, audits might also involve automated data checks, like detecting if any service isn't reporting data properly or if there are anomalies (such as consistently low ridership reported that seems to conflict with independent observations) (Chen et al., 2025).
- **Crisis Management and Intervention:** When unexpected disruptions occur that are beyond the scope of System 2 to handle, System 3 steps in with directives. For instance, during a city-wide emergency (snowstorm, etc.), System 3 might override normal operations to implement an emergency mobility plan (like canceling school buses to divert resources to hospitals). It has the authority to temporarily redirect or reconfigure System 1 activities for the sake of the whole (Jackson, 2000).
- **Interface with Higher Government Levels:** System 3 also communicates upwards with metropolitan/regional or national transport systems (which could be seen as the environment or a higher-recursive system). For example, coordinating with a regional rail service that enters the city, or complying

with national transportation regulations. This is part of managing the "inside" of the system to fit with the "outside" demands (Murphy & Regenold, 2023).

In our mobility system design, System 3 could be embodied by a City Mobility Board or Department that includes not only officials but potentially representation from various stakeholders (private operators, citizen groups, etc. as advisors or members). A multi-stakeholder board can increase transparency and trust – an important consideration since this central function wields significant power. It should be noted that Beer's model is neutral on whether System 3 is one entity or several; the key is that the functions are covered. In some contexts, part of System 3 might be automated (e.g., an algorithm optimizing traffic flows) and part human (e.g., managers deciding on service contracts) (Beer, 1981).

One innovative approach within System 3 to incorporate citizen self-management is to practice **co-management**. For example, in some cities, transit authorities have "community advisory committees" that regularly meet with managers and provide input on operational decisions (like where to extend a bus route). This can be institutionalized so that community voices are part of the System 3 decision-making process. In a more radical self-management angle, one could imagine that some System 1 units (like a cooperative) actually have representation in System 3 – effectively they are both operators and part of the management council. This idea has parallels in cooperative federations where member co-ops collectively govern shared infrastructure (Walker, 1998). It aligns with the democratic ethos of *autogestión*, bringing the managers and the managed closer together.

System 3 in our model must work closely with Systems 4 and 5, as we will see. It handles the present operations but needs to be guided by future planning (System 4) and policy (System 5). A potential pitfall is rigidity – if System 3 overextends control, it could stifle the adaptability of System 1 units or the forward-looking initiatives of System 4. Thus, designing System 3 requires balance: it should intervene enough to ensure integration and efficiency, but not so much as to micromanage or suppress local innovation. The presence of active citizen-run units might actually help check System 3's tendency to centralize, because these units might insist on a degree of independence, thereby forcing System 3 to adopt a facilitative rather than dictatorial style. This can make the whole system more flexible and resilient (Espinosa & Martinez, 2025; Oregi et al., 2025).

## System 4: Strategic Planning, Innovation, and Adaptation

For **System 4** in our city mobility system, we establish a function (or set of functions) dedicated to understanding the changing external environment and steering the system's evolution accordingly. This could be realized as a **Mobility Strategy and Innovation Office** within the city government, or a cross sector task force, or even multiple units (like a research division in the transit agency plus partnerships with local universities). The key aspects of System 4 design include:

- **Environmental Scanning:** System 4 monitors external factors that could impact urban mobility. These include technological trends (rise of autonomous vehicles, new battery technologies for EVs, emergence of Mobility-as-a-Service digital platforms), social trends (e.g., remote work reducing commute demand, aging population needing more accessible transport), economic factors (fuel prices, funding availability), and environmental factors (air quality issues, climate change effects). It might produce regular horizon-scanning reports or foresight studies. For example, System 4 notes that autonomous shuttles are being tested in other cities – it analyzes what that could mean for our city's bus drivers and service patterns in 10 years.
- **Community and Stakeholder Input for Foresight:** A good System 4 doesn't just rely on experts; it also harnesses the collective intelligence of citizens and stakeholders. This can include conducting surveys, holding public forums on future mobility, or running pilot programs to test new concepts with user participation. Engaging citizens in scenario planning workshops for future transport ("how would you feel about converting more street space to bike lanes?" etc.) serves two purposes: it gathers knowledge (citizens often know local challenges best) and it builds public support for eventual changes. In our design, we envision periodic **Mobility Labs or Living Labs** where innovations are co-created. For instance, the city could invite tech companies, transit users, hacktivists, and academics to collaborate on an experimental app that encourages ride-sharing, with the results feeding back into official strategy if successful.
- **Planning and Policy Proposals:** Based on external scanning and internal analysis, System 4 develops strategic plans and policy recommendations. This could be a 5-year city mobility plan that lays out investments (e.g., extending a metro line, introducing congestion pricing) or conceptualizes a shift like "15-minute city" planning (where amenities are within a short walk/ bike of residents). These plans, however, must be reconciled with what System 3 sees as current operational capabilities – this is the famous **3–4 homeostat** in VSM that balances present and future. If System 4 proposes something too far beyond current reality (say, phasing out all cars in 5 years), System 3 will push back, and System 5 might mediate. The design should include formal processes for this reconciliation: e.g., the strategic plan goes for consultation with the operations managers and for public comment before being finalized.

- **Innovation Projects and R&D:** System 4 can initiate pilot projects or experiments to try out new mobility solutions on a small scale. For example, trialing an on-demand minibus service in one district, or deploying some AI-based traffic signal optimization in a few intersections. These provide learning which System 4 analyzes and then, if promising, passes to System 3 to implement more widely. In our design, we allocate some budget specifically for innovation and experimentation, a sandbox where failing is acceptable as long as we learn (a very System 4 approach).
- **External Relationships and Learning:** The System 4 function should also network with other cities and external bodies to learn best practices and bring in knowledge. This could mean attending global conferences, collaborating in international programs (like C40 Cities for climate and transport), and keeping tabs on scientific research in urban mobility. By linking outward, System 4 ensures our city isn't insular; it can import ideas and even jointly develop solutions with others. For instance, if a neighboring city has a successful bike-share scheme, System 4 might study it and then adapt the concept at home.

One novel concept in strategic urban mobility is building *digital twins* of the city for simulation – essentially a virtual model to run scenarios. A System 4 could oversee such a digital twin for mobility, allowing planners and even citizens (through user-friendly interfaces) to simulate what-if situations (White et al., 2021). This enhances the planning process with data-driven insight and could even be used participatorily (imagine a public workshop where people tweak parameters in the simulation to see outcomes for, say, adding a bus lane vs. not).

Now, how do citizens play into System 4 in our design? Apart from the input methods mentioned, one could institutionalize citizen involvement via something like a **Mobility Futures Council** which includes citizen representatives, futurists, etc., advising the city. Another angle is academia and civic tech communities – by funding local universities or start-ups to focus on urban mobility challenges, the city can indirectly engage many citizens in innovation.

Crucially, System 4 must remain connected to System 5 (policy) to ensure that strategic explorations align with the city's identity and values. For example, if the city's ethos (System 5 decision) is to prioritize equity, System 4's proposals should reflect that (like making sure innovations also serve low-income neighborhoods, not just affluent areas).

## System 5: Policy, Identity, and Governance

**System 5** in our city mobility system is about governance at the highest level – deciding the core purpose and values of the mobility system and ensuring that the activities of Systems 3 and 4 (and by extension the whole system) are in line with those values (Beer, 1981; Espejo & Reyes, 2014). In a democratic public sector context, System 5 could be identified with the City Council or Mayor's Office, specifically as it relates to transportation policy. It might also be a Board of Directors if a semi-independent metropolitan transport authority exists. Regardless of form, System 5's design centers on the following:

- **Defining Vision and Mission:** System 5 articulates what the city's mobility system is fundamentally trying to achieve. Is it to provide affordable transport for all? To eliminate traffic fatalities? To achieve carbon neutrality? To support economic growth? Typically, it will be a blend of such goals. For instance, a mission statement could be: "To deliver a safe, sustainable, and inclusive mobility system that connects all residents with opportunities." This high-level policy then filters down as a guiding star for strategies and operations. In practice, such vision might be enshrined in official documents like a City Transportation Master Plan or a Mobility Charter (Oregi et al., 2025; Murphy & Regenold, 2023).
- **Setting High-Level Policies and Standards:** Based on the vision, System 5 sets non-negotiables or key targets. For example, it might establish that at least 50% of trips should be by sustainable modes by 2030, or a policy that any new transportation technology (like autonomous vehicles) must align with public interest. It could also declare standards like a maximum fare for public transit to keep it affordable, etc. These become constraints within which System 3 must operate. A real-world example: many cities have adopted "Vision Zero" (zero road deaths) as a policy – that directly influences priorities in operations. Another example: a policy of participatory decision-making – System 5 might require that any major route change goes through a community consultation process, embedding citizen participation as a rule (Albino, Berardi, & Dangelico, 2015; Nam, 2011).
- **Ensuring Balance Between Present and Future (Mediating 3 and 4):** System 5 receives input from System 3 about operational realities and from System 4 about needed changes and external threats. Sometimes these conflict; System 5's role is to make the ultimate decisions balancing short-term needs with long-term sustainability (Beer, 1985; Jackson, 2000). For instance, System 4 may propose a major overhaul like congestion pricing to reduce car traffic, but System 3 warns it could disrupt current traffic patterns

and anger drivers. System 5 (e.g., the elected officials) weigh the options and set a course (maybe phasing it in with political measures to garner acceptance). In our design, this could take the form of the City Council approving strategic plans or budgets that incorporate both perspectives. The key is that someone has the authority to say "yes, we commit to this path" and thus give System 3 and 4 a coherent directive to follow.

- **Identity and Culture:** While harder to quantify, System 5 also shapes the culture of the mobility system. If citizen self-management is valued, System 5 exemplifies that by, say, celebrating community initiatives publicly, including them in ceremonies or official narratives of the city's achievements. This top-level support can permeate through layers, encouraging a culture of collaboration and respect between agencies and citizen groups. Conversely, System 5 also embodies the public accountability of the system – if things go wrong (say, persistent congestion or a transit failure), it's at this level that the system's legitimacy is maintained or questioned. Thus System 5 must maintain public trust through transparency and responsiveness, which often implies that the governing body stays connected to citizen concerns (possibly through public forums or oversight committees) (Espinosa & Martinez, 2025).
- **Structural Governance and Constitutions:** In designing System 5, we might also consider how formal governance structures could innovate. For example, could there be a participatory budgeting process for transport, where citizens directly vote on a portion of the mobility budget? That would make citizens part of System 5 for that financial decision context. Or perhaps a rule that a certain number of seats on the Transport Authority Board are reserved for citizen representatives or experts from outside government. These design choices can hard-wire citizen influence at the highest level of decision-making for the mobility system (Walker, 1998; Díaz-Nafría & Guarda, 2018).

As an illustrative example, consider a city that declares climate action as part of its identity. System 5 (the city leadership) sets a policy that by 2035 the city will drastically cut transport emissions. This leads to decisions like investing in electric buses, building more bike lanes, and disallowing new highway expansions. System 3 carries those out operationally, System 4 aligns future plans with it (e.g., exploring hydrogen buses as tech emerges), and System 2/1 adapt their day-to-day to support it. Meanwhile, citizens are engaged by that vision – community groups might form to support bike lane development, etc. This positive feedback loop between an inspirational System 5 vision and grassroots energy can greatly reinforce viability, because the populace and the system share a common identity and purpose (Medina, 2011; Oregi et al., 2025).

In a contrasting negative scenario, if System 5 is misaligned with citizen values (e.g., a policy direction that the public largely opposes, such as eliminating a beloved tram line purely for cost reasons), viability is threatened by loss of legitimacy. People might resist or circumvent the system (protests, or just not using services). Hence, designing System 5 also means aligning it with the democratic will – not an easy task as multiple interests exist, but processes like public consultations, referenda, and stakeholder representation can help calibrate System 5's directives to what is broadly acceptable (Pira, 2021; Albino et al., 2015).

## Recursion and the Bigger Picture

It's worth noting that our city mobility system itself sits in a recursive context. It could be **System 1** of a larger system (like a metropolitan or regional transportation system), and it contains System 1 units that themselves have internal viability structures (like a transit agency's internal departments). The viable system design can be applied at those levels too (Beer, 1985; Jackson, 2000). But for clarity, we focused on one level (the city-wide). Recursion means that a neighborhood's local mobility (perhaps managed by a mini-council including citizen members) could be a viable subsystem feeding into the city's viable system, and likewise the city feeds into a nation's transportation viable system (Walker, 1998; Espinosa & Martinez, 2025). Designing with recursion in mind ensures that each level has the appropriate autonomy and resources. For example, if regions or districts in the city are quite distinct, the city might decentralize some System 3 functions to borough-level transportation offices (with their own Systems 2–5 locally) that roll up into the city's System 1 units.

## Information Flows Summary

Now that we have sketched each subsystem, let's summarize the major information and control flows in this VSM-inspired mobility system design, as this is crucial to the model's cybernetic control (Beer, 1981; Espejo & Reyes, 2014):

- From each **System 1 unit** (bus service, metro, community shuttle, etc.), information about operations (e.g., ridership, punctuality, incidents) flows upward to **System 3**. Likewise, commands or resource allocations flow downward from System 3 to the units (e.g., "increase frequency on Route X" or "here's your budget for next quarter"). Additionally, lateral information flows between System 1 units occur via **System 2** (e.g., real-time service coordination messages).

- **System 2** channels convey data horizontally (unit-to-unit) and some distilled data vertically to System 3 as needed for oversight. System 2 likely provides System 3 with synthesized reports rather than every raw detail, filtering noise – this is variety attenuation (Ashby, 2018; Geherson, Santi, & Ratti, 2021).
- **System 3** sends up to **System 5 and 4** information about internal performance and any issues that need policy or strategic input ("Inside-and-Now view"). It also receives from System 5 broad directives (policy) and from System 4 suggestions or alerts about external changes (Espejo & Reyes, 2014).
- **System 3*** (audit) can occasionally bypass normal channels to independently check a System 1 (like an on-site inspection), feeding that info back to System 3 to compare with the reported info (Beer, 1985; Murphy & Regenold, 2023).
- **System 4** gathers external data (from the environment: citizen feedback, tech trends, other cities, etc.) and sends upward to **System 5** analyses of "Outside-and-Future" conditions and possible strategies. It receives from System 5 the approval or refocusing of strategies based on identity/policy, and from System 3 it receives reality checks or current capability info (Albino et al., 2015; Chen et al., 2025).
- **System 5** ultimately issues the cohesive set of policies and identity definitions that trickle down: it might formalize them in a strategic plan (to System 4) and guidelines to System 3 ("we will prioritize X over Y"). It also might engage directly with the public (since System 5 often corresponds to elected officials who communicate with citizens) (Nam, 2011; Pira, 2021).

These flows ensure continuous feedback: operational feedback going up, strategic feedback going down, and adjustment at each level. The system as a whole can learn and adapt because changes in the environment (detected by System 4 or even at System 1 frontline) eventually lead to a response that changes how operations run, closing the loop (Beer, 1981; Espejo & Reyes, 2014). Additionally, because information is ideally shared transparently (e.g., an open data approach), even citizen groups can observe system performance and add their insights, creating a more participatory feedback culture (Oregi et al., 2025; Díaz-Nafría & Guarda, 2018).

By explicitly incorporating citizen-run services and participation channels, we've expanded the traditional VSM to be more **polycentric**. There isn't a single command hierarchy, but a network of semi-autonomous actors with a common governance framework – a model increasingly relevant to smart city paradigms (Albino et al., 2015; Geherson et al., 2021). This is analogous to how some natural systems operate: for instance, the human body has autonomous subsystems like organs that self-regulate, but all under overall coordination via the nervous system and brain – and interestingly, some functions like reflexes happen in a distributed way, akin to local citizen actions, yet still beneficial to the whole (Wiener, 2019).

## DISCUSSION

The theoretical model outlined above promises several potential benefits for urban mobility management, but it also raises practical considerations and challenges. In this section, we reflect on how a VSM-based, citizen-integrated approach might perform and what issues need attention when moving from theory to practice.

### Enhancing Resilience and Adaptability

One of the strongest arguments for applying the VSM to mobility is improved resilience. By having multiple autonomous units (multiple modes, multiple providers, including community initiatives), the system avoids single points of failure. If one subsystem faces a disruption, others can compensate. For example, if a bus drivers' strike occurs, perhaps the city can amplify support for bike-sharing and car-pooling – variety of options cushions the shock (Freund et al., 2020; Díaz-Nafría & Guarda, 2018). The VSM structure, with a dedicated System 4 and feedback loops, makes the system more adaptive to change than a static bureaucratic system. It institutionalizes foresight and learning (Beer, 1981; Espejo & Reyes, 2014). Thus, as mobility technologies or behaviors change, the system can iteratively adjust itself rather than requiring a crisis to force change.

The inclusion of citizens in these loops means adaptability is not purely top-down; new ideas and adaptations can bubble up from the grassroots (such as a new app or service invented by residents) and be recognized and integrated by the higher systems (Oregi et al., 2025). In effect, the design harnesses what some call "collective intelligence" for city mobility, not relying only on a central brain but on a network of senses and smaller brains – an embodiment of distributed control that cybernetics often advocates for complex systems (Ashby, 2018; Geherson, Santi, & Ratti, 2021).

### Citizen Engagement and Social Sustainability

By formally recognizing citizen contributions (in System 1 and advisory roles in Systems 3–5), the mobility system can achieve greater public buy-in and legitimacy (Nam, 2011; Walker, 1998). When people feel they have agency and voice in mobility decisions, they are more likely to support and use the services available. This addresses a common failure in top-down planning: lack of usage or acceptance. For instance, infrastructure like bike lanes or transit lines sometimes fail simply because the community was not consulted and thus did not desire or trust them. Our model's emphasis on participatory governance could mitigate such issues (Albino et al., 2015; Oregi et al., 2025).

Additionally, citizen involvement can improve equity outcomes. Residents from underserved neighborhoods can highlight needs that might otherwise be overlooked. A viable system that includes their voice is more likely to distribute services fairly rather than concentrate on commercially profitable routes or zones alone (Espinosa & Martinez, 2025; Pira, 2021).

## Complexity Management

Urban mobility is a quintessential complex system – with a multitude of agents, unpredictable interactions, and continuous flux. The VSM is explicitly a tool for managing complexity by matching it with a multi-layered control system (requisite variety principle) (Beer, 1985; Jackson, 2000). A city that tries to manage everything centrally (without VSM decentralization) may be overwhelmed by complexity and fall into inefficiencies or chaotic oscillations. Conversely, a city that is too decentralized (no integrating meta-system) can have fragmentation and incoherence (each actor optimizing locally but the whole suffers – e.g., too many cars because each individual chooses it, resulting in collective congestion).

The VSM-based design strikes a balance, providing enough central integration but not stifling local variety (Espejo & Reyes, 2014; Murphy & Regenold, 2023). In effect, it offers a structured way to apply Ashby's law: the variety of travel options and management responses (via multiple System 1s and an active System 4) meets the variety of citizen needs and external changes, while the coordinating System 2 and regulating Systems 3/5 attenuate the variety to manageable levels for decision-making (Ashby, 2018; Beer, 1981).

## Compatibility with Existing Institutional Structures

Many cities are bound by legal frameworks that separate responsibilities (e.g., traffic management vs public transit might belong to different departments, or community initiatives might not have legal standing). Implementing a VSM model may require institutional innovation, such as creating a unified Transport Authority or forming formal partnerships with non-government providers (Oregi et al., 2025; Espejo & Reyes, 2014). It might also run into labor issues (e.g., transit unions might be wary of volunteer efforts being seen as taking jobs, so they must be assured roles are complementary). These institutional and political aspects would need careful negotiation. Pilot projects and Memorandums of Understanding can gradually build a cooperative environment (Nam, 2011).

## Measuring Success

To validate a viable mobility system, one needs metrics beyond the traditional (ridership, cost). We'd need to measure things like responsiveness (how quickly can the system adapt to a route disruption?), inclusion (are more diverse groups served?), sustainability (emissions, etc.), and perhaps even the satisfaction of participating citizens. The VSM doesn't provide those metrics, but it provides a framework that likely improves them if done right (Beer, 1985). Over time, adjusting the model based on such measurements would itself be a task of **System 4** feeding into **System 5** decisions – meta-learning about how the system is working (Espinosa & Walker, 2011).

## Overall Vision

While challenging, the integration of VSM with citizen participation offers a compelling vision: a learning, evolving mobility system that treats the community not just as customers but as co-designers and co-managers. This could accelerate innovation (since ideas can come from anywhere in the system) and build a stronger social contract (people support what they help create). It resonates with trends in governance toward more collaborative networks rather than strict hierarchies (Bollier, 2016; Rodríguez-Oroz et al., 2019).

It is interesting to note that some aspects of this model are already emerging in various forms around the world. For instance, "Mobility as a Service" (MaaS) platforms, which integrate different transport modes into one app, are a kind of **System 2/3** combination in a digital shell; they coordinate multiple services and present them as one system to the user (Geherson, Santi, & Ratti, 2021). However, many MaaS initiatives are industry-led; our approach would be to put them in a public/civic context and ensure they incorporate citizen governance. Similarly, cities are experimenting with participatory budgeting including transport projects (Cabannes, 2015), and with open data portals that let civic hackers improve city services (Pira, 2021). These can be seen as fragments of a viable system approach that need to be knitted together under a holistic model.

## On Critiques of the VSM

One critique that has been raised of the VSM in general is that it can be seen as too abstract or idealized for messy real-world situations (Jackson, 2000). We acknowledge that our description here is idealized to an extent – reality will likely be more ad hoc. However, we view the VSM not as a rigid blueprint but as a guiding heuristic (Beer, 1985; Espejo & Reyes, 2014). Cities can adopt the principles (like distributed autonomy with coordination, or structured feedback channels) in ways that fit their context. The virtue of the VSM is that it reminds planners to cover all necessary functions (so you don't forget, say, to have an intelligence-gathering mechanism for future changes, which many traditional organizations neglect). Even if not implemented in full fidelity, using VSM as a checklist for capabilities can strengthen a mobility system's design (Espinosa & Martinez, 2025).

## CONCLUSION

This chapter explored how Stafford Beer's Viable System Model (VSM), a foundational framework from management cybernetics, can be applied to the design of an urban mobility system, particularly by integrating citizen self-management. We began by outlining the complex challenges facing contemporary urban transportation and the limitations of top-down, siloed solutions. In contrast, the VSM offers a recursive and adaptive architecture, identifying five necessary subsystems—Operations (System 1), Coordination (System 2), Control (System 3), Intelligence (System 4), and Policy (System 5)—that together sustain systemic viability in changing environments.

Building on this model, we proposed a theoretical design where diverse transport actors—including public agencies, private providers, and community-led services—operate as semi-autonomous System 1 units. Their interactions are stabilized through System 2 mechanisms such as ICT platforms and shared protocols, while System 3 entities (e.g., a City Mobility Authority) ensure resource optimization and regulatory coherence. System 4 functions provide foresight and strategic responsiveness, and System 5 embodies the identity and core values of the mobility system, aligning it with democratic mandates and long-term societal goals. Importantly, the model embeds citizen participation across all levels—not only as service operators but also as contributors to planning, feedback, and decision-making processes.

The VSM's cybernetic logic provides several advantages. It enhances adaptability and resilience through distributed intelligence and recursive organization. By satisfying Ashby's Law of Requisite Variety, the system can better match the diversity of user needs and environmental challenges. Citizen engagement strengthens legitimacy and encourages more equitable and user-centered solutions. Moreover, the design fosters a learning system: capable of sensing, adjusting, and evolving over time based on both internal performance and external signals.

Nonetheless, transitioning from theory to practice presents significant challenges. Effective implementation requires breaking down institutional silos, building interoperable information systems, and fostering a culture of transparency and collaboration. Coordination burdens may grow as more actors are included, and governance arrangements must evolve to balance autonomy with accountability.

Conflicts between stakeholders—especially between formal institutions and informal or volunteer initiatives—must be anticipated and addressed through clear procedural mechanisms and inclusive governance design.

Despite these difficulties, the changing landscape of urban mobility presents an unprecedented opportunity. The rise of new mobility services, the proliferation of digital data, and growing civic engagement create fertile ground for cyber-systemic innovation. The VSM need not be applied in a rigid or totalizing way; rather, it offers a flexible design heuristic. Cities can begin with localized pilot projects, integrating community initiatives within formal systems, and iteratively expanding the viable structure as capacity and trust grow.

In conclusion, a VSM-inspired approach reframes urban mobility not as the management of discrete assets, but as the dynamic governance of interrelated actors, flows, and communications. It envisions a transport system that is not only technically efficient, but also socially sustainable, politically legitimate, and structurally adaptable. In this sense, the city becomes a living system—capable of learning, evolving, and co-creating its future with its citizens. This cybernetic perspective offers a powerful orientation for planners, system designers, and civic leaders seeking to build mobility systems that are resilient, inclusive, and truly viable over time.

## REFERENCES


Albino, V., Berardi, U., & Dangelico, R. M. (2015). Smart Cities: Definitions, Dimensions, Performance, and Initiatives. *Journal of Urban Technology*, *22*(1), 3–21. DOI: 10.1080/10630732.2014.942092

Ashby, W. R. (2018). *An Introduction to Cybernetics*. Creative Media Partners, LLC.

Beer, S. (1979). *The Heart of Enterprise*. Wiley.

Beer, S. (1981). *Brain of the firm: the managerial cybernetics of organization*. Wiley.

Beer, S. (1985). *Diagnosing the System for Organizations*. Wiley.

Chen, Y., Li, C., Wang, W., Zang, Y., Chen, X. M., & Gao, Z. (2025). The landscape, trends, challenges, and opportunities of sustainable mobility and transport. *npj Sustainable Mobility and Transport, 2*(8). DOI: 10.1038/s44333-025-00026-8

Debnath, A. K., Chin, H. C., Haque, M. M., & Yuen, B. (2014). A METHODOLOGICAL FRAMEWORK FOR BENCHMARKING SMART TRANSPORT CITIES. *Cities (London, England)*, *37*, 47–56. https://vuir.vu.edu.au/34171/. DOI: 10.1016/j.cities.2013.11.004

Díaz-Nafría, J. M., & Guarda, T. (2018). Participatory Sensing in Sustainable Mobility: biciLAB Model. In Á. Rocha & T. Guarda (Eds.), *Developments and Advances in Defense and Security: Proceedings of the Multidisciplinary International Conference of Research Applied to Defense and Security (MICRADS 2018)* (Vol. 94, pp. 97-104). Springer International Publishing. DOI: 10.1007/978-3-319-78605-6_8

Espejo, R., & Reyes, A. (2014). *Organizational Systems: Managing Complexity with the Viable System Model*. Springer Berlin Heidelberg.

Espinosa, A., & Martinez, A. (2025). The Viable System Model to Support Sustainable SelfGovernance in Communities: Learning from Case Studies. *Systemic Practice and Action Research*, *38*(14), 14. Advance online publication. DOI: 10.1007/s11213-025-09724-3

Freund, K., Bayne, A., Beck, L., Siegfried, A., Warren, J., Nadel, T., & Natajaran, A. (2020). Characteristics of ride share services for older adults in the United States. *Journal of Safety Research, 72*, 9-19. DOI: 10.1016/j.jsr.2019.12.008

Gallego, S., Rodriguez, R., & García, M. (2018). Design of a Conceptual Model for Improving Company Performance Based on Lean Management Applying the Viable System Model (VSM). *European Journal of Engineering and Formal Sciences*, *2*(2), 49–63. DOI: 10.26417/ejef.v2i2.p50-64

Geherson, C., Santi, P., & Ratti, C. (2021). Cybernetic cities: designing and controlling adaptive and robust urban systems. In Portugali, J. (Ed.), *Handbook on Cities and Complexity* (pp. 195–208). Edward Elgar Publishing., DOI: 10.4337/9781789900125.00018

Hosford, K., Pitman, B., & Winter, M. (2021). Facilitators and barriers to the implementation of community-based transportation services for older adults: Evidence from six case studies. *Transportation Research Interdisciplinary Perspectives, 24*(101062), 101062. Advance online publication. DOI: 10.1016/j.trip.2024.101062

Jackson, M. C. (2000). *Systems Approaches to Management.* Springer US.

Johnson, A. M., Zhou, C., Haviland, M., & Mendoza, J. A. (2024). Evaluation of a walking school bus program: A cluster randomized controlled trial. *The International Journal of Behavioral Nutrition and Physical Activity, 21*(55), 55. Advance online publication. DOI: 10.1186/s12966-024-01602-w PMID: 38730407

Medina, E. (2011). *Cybernetic Revolutionaries: Technology and Politics in Allende's Chile.* MIT Press. DOI: 10.7551/mitpress/8417.001.0001

Murphy, S., & Regenold, T. (2023). Applying the Viable System Model to Local Government in a Post-Pandemic Context. In Azukas, M. E., & Kim, M. (Eds.), *Reimagining Systems Thinking in a Post-pandemic World* (pp. 79–96). IGI Global., DOI: 10.4018/978-1-6684-7285-9.ch004

Nam, T. (2011). Conceptualizing smart city with dimensions of technology, people, and institutions. *Proceedings of the 12th Annual International Conference on Digital Government Research.* DOI: 10.1145/2037556.2037602

Oregi, I. L., Urra-Uriarte, S., Gonzalez, I., Anaya, M., & Molina-Costa, P. (2025). Enhancing Citizen Participation in Citizen-Centered Smart Cities: Insights from Two European Case Studies. *Urban Science (Basel, Switzerland), 9*(5), 140. DOI: 10.3390/urbansci9050140

Pira, M. (2021). A novel taxonomy of smart sustainable city indicators. *Humanities & Social Sciences Communications, 8*(197), 197. Advance online publication. DOI: 10.1057/s41599-021-00879-7

Walker, J. (1998). *The Viable Systems Model a guide for co-operatives and federations.* ICOM (Industrial Common Ownership Movement). https://metaphorum.org/wp-content/uploads/2022/11/vsm-coop-walker.pdf

Wiener, N. (2019). *Cybernetics Or Control and Communication in the Animal and the Machine.* MIT Press., DOI: 10.7551/mitpress/11810.001.0001

## KEY TERMS AND DEFINITIONS

**Adaptive Capacity:** The ability of a system to adjust its functioning in response to changes, disturbances, or new information. A high adaptive capacity in a city mobility system means it can reroute services, incorporate new transport modes, or change policies relatively quickly and smoothly as conditions evolve. This is related to System 4 and the feedback processes in the VSM, enabling continuous learning and adjustment.

**Citizen Self-Management (Autogestión):** A form of governance or service provision where users or community members actively manage or co-manage resources and services, rather than relying solely on centralized authorities. In urban mobility, this can refer to community-run transport services or citizen involvement in decision-making and operations.

**Cybernetics:** The science of communication and control in systems, whether machines, organisms, or organizations. Key concepts include feedback loops, self-regulation, and adaptation. In this context, cybernetics provides a lens to design city mobility systems that can self-correct and respond to changes through feedback mechanisms.

**Feedback Loop:** A fundamental concept in systems theory and cybernetics where outputs of a system are fed back as inputs, forming a loop that can stabilize or change the system's behavior. A feedback loop in urban mobility might be, for example, using real-time traffic data (output of the system) to adjust signal timings (input into the system) to alleviate congestion.

**Participatory Governance:** Governance processes that actively involve stakeholders, especially the public or community groups, in decision-making. For mobility, participatory governance means citizens have channels to influence policies, plans, and even day-to-day operational decisions, ensuring the system reflects their needs and values.

**Requisite Variety:** A principle from cybernetics (formulated by W. Ross Ashby) stating that to effectively control a system, a regulator must have as much variety (responses, options) as the system it aims to control exhibits. In practice, a city mobility system must offer a sufficient range of options and responses to handle the diversity of travel demands and disruptions (e.g., multiple transport modes to handle different scenarios).

**Smart City:** A city that leverages technology, data, and innovative governance to improve the quality of life for its citizens, often emphasizing sustainability, efficiency in services like transportation, and participatory approaches. Smart mobility is a core component, using ICT (information and communication technology) to integrate and enhance urban transport.

**Systemic Coordination:** The organized synchronization of various parts of a system to ensure they work together without conflict. In a mobility context, systemic coordination could refer to aligning schedules between different transit modes, sharing data among agencies, or collaborative planning across departments: essentially the System 2 function in VSM.

**Urban Mobility System:** The interconnected network of transportation modes, infrastructure, people, and institutions that facilitate movement in a city. It includes public transit, private vehicles, non motorized transport (walking, cycling), and emerging modes, as well as management structures and policies governing these elements.

**Viable System Model (VSM):** A theoretical model of organizational structure developed by Stafford Beer in cybernetics. It describes the necessary subsystems (Systems 1 through 5) and their interactions for any complex system to be viable (capable of independent existence and self-management). The VSM emphasizes balanced autonomy and control through operations, coordination, control, intelligence, and policy functions.

# Chapter 8
# Smart Cities and Infrastructure Industry

**R. Velmurugan**
https://orcid.org/0000-0002-7925-9757
*Karpagam Academy of Higher Education, Coimbatore, India*

**J. Sudarvel**
https://orcid.org/0000-0001-6656-1992
*Karpagam Academy of Higher Education, Coimbatore, India*

**K. Jothi**
https://orcid.org/0000-0002-8625-863X
*KPR College of Arts Science and Research, Coimbatore, India*

**Ravi Thirumalaisamy**
https://orcid.org/0000-0002-5954-5940
*Modern College of Business and Science, Oman*

## ABSTRACT

*The rapid urbanization has led to the emergence of smart cities that are developed so as to ensure sustainability and efficiency by way of integrated technologies. The infrastructure, mobility, energy, safety, and resources are all optimized with help of IoT, AI, and data analytics. Real time monitoring as well as prediction maintenance is made possible through smart grids, intelligent transport, BIM, digital twins, automated utilities. The innovations influence improved service delivery, lower emissions and green development. Nevertheless, inclusive and resilient urban futures can be attained, given strong governance and collaboration challenges, such as cybersecurity, data privacy, technological integration.Keywords: Smart cities, infrastructure development, sustainable urbanization, digital transformation, IoT integration*

DOI: 10.4018/979-8-3373-2802-7.ch008

## INTRODUCTION

The challenges that urbanization has placed across the globe have been accelerated beyond anyone's imagination and these need innovative and sustainable solutions. As of now, nearly 55 percent of the world's population lives in urban areas, and this is expected to increase to 68 percent by 2050; (United Nations, 2018). To this rapid urban expansion comes extreme pressure on existing urban infrastructure, resource, goverance mechanisms. Not only does urbanization expand housing demand, transportation, energy, water supply, traffic congestion, pollution, waste management and socio-economic inequality, but it also presents intricate multi-disciplinary problems in housing, energy, and transportation disciplines. Given these manifold challenges, the idea of 'the smart city' has evolved as a new strategic idea to solve the urban landscape issues by merging state of the art technology with sustainability concepts.

A smart city is an urban ecosystem that makes use of digital technologies to achieve high efficiency and performance in urban services and thus improving urban quality of life. Despite the ubiquity, little data exists currently on the implementation of smart cities in the developed countries, but it is generally believed that the core aspects typically include the deployment of the Internet of Things (IoT), Artificial Intelligence (AI), big data analytics and sensor based infrastructure for improving operations such as traffic management, energy consumption, waste disposal and public safety (Batty et al., 2012; Albino et al., 2015). These technologies enable real time monitoring, predictive analytics and data driven decision making that enables the city administrators to understand city urban environment and govern it in a more effective manner.

Smart city technologies have been especially successful in various leading urban centers across the world with most recently adopting them. One example of this is that Singapore is frequently mentioned as an example of a smart city development. Singapore, which launched its “Smart Nation” initiative in 2014 through the biggest tech, for instance, autonomous vehicles, pay as you go payment, AI surveillance system as well as a national sensor platform to make its urban environment extremely connected and effective (Smart Nation Singapore, 2020). A notable project in this regard is its involvement of AI and IoT in its water management system that monitors the leaks, tracks the water consumption patterns, and predicts the maintenance needs to save on both the expenses and the environmental concerns.

Like Barcelona, Europe has also had many smart city strategies piloted by the city, beginning with Barcelona. They have developed an omnibus urban platform referred to as ‘Sentilo’ that encompasses coding data from hundreds of sensors deployed across the city that measure the the ambient noise and the level of air pollution, availability of parking spots or illumination of the streets (Bakıcı et al., 2013). This smart lighting system in the city has become so bright, that it adjusts the brightness to the extent of pedestrian's movement and traffic density, thereby significantly cutting down on its energy consumption and boosting public safety. Barcelona's smart ideas, reported by the European Commission (2017), have ended up with a 30 percent reduction on water consumption and a 47 percent better capacity on traffic flow in selected areas.

Further, as it makes decision making happen in an autonomous and predictive manner, the integration of AI makes the scope of smart cities much bigger. These arealgoirthms can be used to machines the forecast of energy demand, the optimization of public trasportation routes and the improvement of the security surveillence systemes. In New York City, the 'Domain Awareness System' integrates over 9,000 CCTV cameras and sensors that use AI for detection of suspicious behavior and support crime prevention (Brantingham et al. 2018). These systems are not only about increasing the urban public governance efficiency, but they also encourage much more proactive urban governance.

Secondly, big data analytics is one of the big pillars of the smart city framework where there is a huge amount of data in which we have to transform that as an underlying insight for us to make use. On the other hand, the data on real time transport of cities such as London are used by both the users and the operators at a timely manner to improve the efficiency in the current Tube and bus systems. Shown in (Hashem et al. (2016)) is the use of data analytics to predict maintenance needs and improve reliability of service, in the context of Transport for London (TfL).

Although these are all waves towards the smart city idea, it is not without its detractors. Scholars and policy makers (Kitchin, 2016) are often preoccupied with issues around data privacy, cybersecurity, digital exclusion and such down the line as ethical AI deployment. Smart cities advocates say that they offer the efficiency, sustainability and efficiency that policy-makers seek, while also claiming that this forms of urbanism are not much more than surveillance extended environments in which personal freedoms could be eroded. Solving for these concerns requires transparent services framework for data governance, equal access to digital infrastructure and public–private partnerships for the advance of technological solutions that will be enablers for all urban population segments.

In addition, building smart cities is costly and infrastructure and financial demands for building smart cities are very high, especially for developing countries. While cities in India have made progress in digital governance and mobility as part of the Smart Cities' Mission initiated in 2015, scalability of its impact is constrained by budgetary restrictions and limited literacy around digital matters within the urban poor communities (Ministry of Housing and Urban Affairs, Government of India, 2022).

The concept of smart cities is thus journeyed into a comprehensive strategy to handle the growing issues of urbanization. Smart cities become a guideline for the development of resilient, inclusive and ready for the future cities by incorporating technology into the process of sustainable urban planning. Interrogation of data not only makes the city more productive but also helps city administrations predict challenges and proactively respond to them. More cities are coming to realize that as they progress down their smart transformation pathway, so does the demand to think inclusively, be highly security conscious, and implement participatory governance. While smart cities represent only technological evolution, it is rather an enormous shift in the manner a city is conceived, is governed and is experienced.

## THE ROLE OF TECHNOLOGY IN URBAN TRANSFORMATION

With the proliferation of extra connectivity between people, data, and things, cities are becoming increasingly smart and intelligent by using connected technologies to turn into the seamless, efficient, responsive and intelligent urban ecosystems. The city of the future will be instituted by means other than the traditional infrastructure and governance models, which are no longer sufficient to cope with the growing demand for services, energy, mobility, and security increasingly required by the high speed of urbanization. Thus, municipalities are integrating smart technologies especially the Internet of Things (IoT), Artificial Intelligence (AI), and data analytics through the cloud into the urban development in sustainable and adaptive way (Hashem et al., 2016; Batty et al., 2012). These tools are helping cities make the transition from reactive to proactive service delivery by enabling cities to better monitor and predict how a service is to be used, to maximize the impact of the resources available, and thus inform better, more agile decisions.

Data collection from such an extensive set of touchpoints across the urban realm is a result of IoT sensors. Among them was traffic signals, streetlights, water pipelines, waste bins, and environmental stations. For example, sensors are widely deployed in Amsterdam's Smart City project to control the speed of cars in and out of the city, regulate traffic lights to suit pedestrian movement, and monitor the real time use of energy (van Winden et al., 2016). Barcelona also leverages an immense IoT sensor network powered through 'Sentilo' platform for gathering data on the air quality, noise level, parking availability, and so on to enhance urban efficiency and reduce its energy consumption (Bakıcı et al., 2013).

Additionally, Artificial Intelligence helps the smart cities gain the extra benefit by processing sensor generated data and deriving actionable insights. It creates automation for change in traffic light, energy demand prediction, and security monitoring (e.g., detecting unusual activity for public safety monitoring) using AI powered platforms. At present, an AI based Domain Awareness System from Microsoft and collaboration with NYPD is used in New York City to help analyze data collected from surveillance cameras, license plate readers, and other sources, and to assist with law enforcement and emergency response (Brantingham et al., 2018).

This technological ecosystem can only be deployed using cloud computing, big data analytics infrastructure. The data coming from this vast number of sensors provided by these platforms gives cities the ability to put away multicore data, run sophisticated analytics, that allow them to forecast trends and help drive long term policy planning. Real time analytics is employed by London's Transport for London (TfL) to manage its large public transportation system to improve bus schedules and minimize service disruptions (Hashem et al., 2016).

Scaling and scaleless, adaptable to the changing needs of the modern urban life. Pioneered by connected technologies, cities are being redefined in tangible ways by the ways in which they operate and relate to the residents that live there reducing environmental footprint and energy consumption, achieving public safety and increasing engagement with their citizen base. Despite that, as Kitchin (2016) points out, there is also a concern regarding this implementation of systems of this sort and of how it impacts data privacy, surveillance and governance ethics, which requires for transparent and inclusive policies for urban tech.

Therefore, smart technologies are not futuristic but vital elements of modern urban planning. IoT, AI, and cloud computing made cities more responsive, resilient, and sustainable by means of the strategic deployment. And as cities evolve into a much more urban condition, these tools will need to be effectively integrated into our cities in order to have cities that are not only intelligent, but levy and livable.

## INFRASTRUCTURE AS THE CORNERSTONE OF SMART CITIES

The creation of smart cities necessitates the presence of the foundation infrastructure that provides the physical and technological elements neither of which can be sustained without one another. However, without a well developed infrastructure, the vision of what smart urbanism is —sustainability, efficiency, and citizen centric services can not be fully realized. Modern smart city infrastructure interoperates digital technology and traditional systems to fashion urban settings that run efficient and less vulnerable still and serve real time conditions and future expectations (Neirotti et al., 2014).

Smart cities have the most prominent components of infrastructure smart transportation systems. And these are systems that use part of sensors, GPS tracking, data analytics and AI to oversee traffic, reduce congestion and optimize public transport. One such leading example is Singapore's Intelligent Transport System (ITS) that has sensors and surveillance systems that monitor traffic patterns real time and adopt dynamic pricing in road tolls as well as manipulate dynamic traffic signals

(Smart Nation Singapore, 2020). These systems have been integrated in such a way to reduce average travel times and enhance commuter satisfaction.

One more important element of the infrastructure in smart cities is energy efficient power grids, usually termed smart grids. Real time electricity distribution and identification of outages, and supply demand balancing all improve energy efficiency and make it possible to integrate renewable sources of electricity. This is an example of the 'Pecan Street Project' in Austin, Texas where smart meters and home energy management systems were introduced so that users can see how much it costs to consume or the utility company optimizes the grid (Pecan Street Inc., 2019). These initiatives lead to the emission of less greenhouse gas and encourage the use of energy economically.

Now more and more urban infrastructure embraces climate responsive building to help with environmental sustainability. These include such things as using energy efficient materials, passive solar design, green roof, and sensor driven HVAC (heating ventilating and air conditioning), which adjusts the temperature according to occupancy. As in Dubai's Sustainable City, buildings are designed to meet the net zero energy challenge through rooftop solar panels, automated systems for water conservations and indoor climate control (Alyami & Rezgui, 2012).

This is possible because embedded sensors are placed in public utility such as roads, bridges and pipelines to ensure continuous condition monitoring and real-time data collection. These sensors detect vibrations, structural strain, temperature changes, material fatigue in order to allow authorities to detect problems before they become major. Sensors are installed in major bridges in South Korea by the Seoul Metropolitan Government to find structural weakness and warn of repairs that would increase safety and cut future maintenance costs (Lee et al., 2018).

Additionally, the predictive maintenance systems based on AI are able to extend the lifecycle of infrastructure assets. These systems take in historical and real time data and predict when such equipment is likely to fail to allow more effective planning of the time slots for the maintenance and repair of the systems. Predictive analytics that were used in London's Crossrail project enabled engineers to monitor tunnel boring machines and rail systems in real time and diminish downtime and enhance productivity (Crossrail Ltd., 2020).

To summarize, this study argues that the smart city design does not rely on a bare physical infrastructure alone, but instead, involves the use of data as a dynamic part contributing to the infrastructure, and subsequently, associates with the smart city framework. Cities can make them sustainable, reactive and future ready by introducing intelligence in roads, buildings, power systems and transportation networks. Massive bandwidth and cost will diminish the need to drive smart city development. But infrastructure will continue to remain the corner stone of smart city development both in terms of improving operational excellence and enhanced quality of life in the urban environment.

## DIGITAL ENGINEERING AND VISUALIZATION TOOLS

Building Information Modeling (BIM) and Digital Twin have greatly shaped the metropolis planning, urban infrastructures management, and urban sustainable development. Technology is the great enabler in terms of data driven planning, precision construction and so much more of the lifecycle for the urban asset. BIM is a digital representation of the physical and functional characteristics of a building site or infrastructure used for the purpose of collaborative platform including architects, engineers, planners and stakeholders. It improves the design accuracy, and at the same time helps discovering conflicts early, allowing comprehensive risk assessment, thus saving resources and meeting the sustainability goals (Eastman et al., 2011).

BIM offers one of the key advantages of building performance simulation and optimization prior to construction commencing. For example, BIM was used to digitize over 14,000 building components within the Sydney Opera House's maintenance model in order to plan for predictive and hopefully lower operational costs (Teicholz, 2013). BIM has been similarly useful in reducing errors, time delays and cost in the UK's Crossrail project, for instance (Crossrail Ltd., 2020).

BIM goes over and above that because it is the real time digital replica of physical city infrastructure. Digital Twin is built based on data from IoT sensors integrated across physical assets; thus, it is updated in a dynamic manner that is different from static BIM models. This makes city plannings and administrators able to simulate these scenarios, to monitor their performance, and to react in time to environmental changes of the urban environment. Traffic congestion, for environmental quality and energy efficiency, are all very valuable problems to solve with Digital Twins.

The city of Singapore itself is a good real world example which has created its own Virtual Singapore National Digital Twin called "Virtual Singapore?" The data is integrated from transport, utilities, environmental and demographic sensors, and advanced simulations can be conducted through such aspects as modelling of heatwaves, optimization of/emergency evacuation routes, or exploration of the impact of high rise developments on wind flow (Smart Nation Singapore, 2020).

The Helsinki city has also applied Digital Twins as they created a real time 3D model based on which new infrastructure planning is made possible along with assessment of environmental impacts and interacting citizens through immersive visualization tools (City of Helsinki, 2020). This has showed the potential of Digital Twin tool to facilitate proactive governance and smarter urban decision making. Basically, BIM and Digital Twins are emerging to change urban planning by connecting physical assets with their digital twins to usher new age of smart, sustainable and responsive city development.

## SMART MOBILITY AND SUSTAINABLE TRANSPORT

Smart mobility, which reimagines the transportation of urban areas through an incorporation of smart traffic system, automated transit solutions and electric vehicle (EV) infrastructure is one of the hallmark of smart cities. With more and more people moving into cities in the midst of rising vehicular emissions, congested roadways, smart mobility becomes a sustainable and sane way to accommodate urban transport needs. With Artificial Intelligence (AI), Internet of Things (IoT) technologies, smart mobility offers better accessibility, less bottlenecks traffic and contributes greatly into environmental and social sustainability (Guerra & Morris, 2018).

Intelligent Transportation Systems (ITS) are at the core of smart mobility initiatives. Such systems utilize sensors, GPS devices and communication networks to gather traffic data which is then analyzed to enhance the effects of traffic signals, plan routes on the fly and also more efficiently manage incidents. For instance, Adaptive Traffic Control System (ATCS) is used by city of Los Angeles which monitors the traffic data in real time to reduce travel delays and fuel consumption (FHWA, 2020). Similarly, the congestion pricing system implemented in London does the same in charging vehicles entering high traffic areas during peak hours, thus leading to less vehicle usage and lower emissions in the city center (Leape, 2006).

The second transformative change in the smart mobility is AI powered route optimization. Crowd sourced traffic data and AI algorithms is used by applications like Google Maps and Waze to suggest the fastest routes, hence avoiding the worst area to bring in traffic, and the longest time to be idling. The Singapore's Land Transport Authority (LTA) uses predictive analytics to collect indicators across all bus services in Singapore so that the readiness in servicing commuters can be assessed and bus frequencies can be adjusted accordingly to ensure bus service reliability and passenger satisfaction (Smart Nation Singapore, 2020).

Smart mobility is also heavily focused on the growth of EVs and supporting infrastructure. EV helps in cutting down of reliance on fossil fuels and helps to generate cleaner air in urban areas. As of 2022 (IEA, 2023), 36 percent of Oslo, Norway's new cars are electric with incentives like free parking, no toll charges and access to bus lanes. Furthermore, smart EV charging stations with smart roll out, that automatically adjust power usage according to the grid demand for energy efficiency and grid stability, are rolled out.

Another innovation of urban mobility is represented by intelligent parking solutions. Using IoT sensors, these systems detect available parking spots and through the use of the mobile apps, they direct drivers to the nearest empty spot. For instance, the smart meter/sensor implementation of the SFpark project in San Francisco helped reduce cruising time as well as reduce vehicle emissions by up to 30% (Pierce & Shoup, 2013).

Systems of automated and autonomous transit are being increasingly explored as efficient and cost effective systems. Driverless pods used in dedicated tracks in cities like Masdar City, UAE, also exist and carry zero emission transport in the urban centers. The "RoboBusLine" pilot project located in Helsinki, was a test of self driving minibuses to serve as additional public transport to solve last miles gap issue (Ammad et al. 2020). Overall, smart mobility is an indispensable part of smart city platform enabling the better efficiency, accessibility as well as sustainability within the transportation systems. Deploying data driven and environmentally friendly solutions will help cities in reducing the congestion, lowering the emissions and provide them better connectivity to improve overall urban living experience.

## SMART GRIDS AND ENERGY MANAGEMENT

The architecture of smart cities significantly depends on energy management and has direct impact on the sustainability, operational efficiency, and environmental impact. With ever growing urban populations and growing energy demands, cities need to adopt intelligent solutions in energy such that supply is consistent and energy conservation and use of renewable energy is promoted. Modern energy management systems are Sound smart grid which use different digital communication technologies and analytics based real time data. Through these grids, the electricity flow is dynamically controlled and a demand response is enabled along with fault detection and load balancing at its optimum for the utility networks (Gungor et al., 2011).

The conventional centralized energy distribution model is significantly different from the smart grids. They enable a two way communication between power producers and the power users and hence aid in integrating decentralized renewable energy sources such as solar panels and wind turbines. Especially, it is important in cutting the reliance on fossil fuels and achieving long term sustainability goals (Fang et al., 2012). One example of this kind of event is the implementation of a smart grid including residential solar panels, energy storage systems, and electric vehicles (EVs) and then balancing supply and demand in real time using that smart grid; that is, the "Smart Energy City" in Amsterdam, in particular. It is circular energy economy; residents can sell excess solar power back to the grid.

A third feature of the smart grids is fault detection. The utility providers can use sensor networks and real time monitoring to locate outages or anomalies quickly, to act before the outages or anomalies extend to a widespread issue that would cause downtime. Examples for use of smart grid technology include the reduction of outage durations and improvements in service reliability as in the case of Austin Energy in Texas, where substations are monitored and failures are detected before they affect consumers (Pecan Street Inc., 2019).

Smart cities also include smart building level energy systems such as intelligent lighting, heating, ventilation, and air conditioning (HVAC) controls and energy monitoring items. IoT sensors and AI are used for tracking occupancy, temperature and energy consumption patterns to maximize energy use through these systems. An example would be that of Barcelona where the smart street lighting that adjusts its brightness by depending on motion of the pedestrians and time of day succeeds energy savings of more than 30% (Bakıcı et al., 2013). Smart HVAC systems can greatly reduce energy costs of commercial buildings by dynamic adjustments of settings in accordance with real time occupancy data, weather forecast at the areas and the state of indoor air quality.

Moreover, energy monitoring platforms contribute to raising awareness and making smart energy consumption more and more appealing. In Helsinki, Kalasatama Smart District allows residents to trace their real energy usage in time and change their behavior effectively leading to lower consumption and smaller environmental footprint (City of Helsinki, 2020).

Additionally, integration of smart metering infrastructure assists in accurate bills, and reduces manual errors on the part of the consumers. Dynamic pricing models (IEA, 2020) are a realized concept in Italy where almost 35 million customers got benefited from a smart metering system rolled out across the country.

Consequently, smart city success relies on smart energy management systems. So, through smart grids, automated building systems, real time energy analytics, cities can incrementally save energy, reduce carbon emission and offer citizens participation towards sustainability in energies. These seem to handle urban energy problems and build more resilient and flexible urban environments.

## Enhancing Public Safety with AI

Public safety is the first pillar in the construction of a smart city and takes a huge leap in the direction of public safety through the incorporation of Artificial Intelligence (AI), facial recognition technologies, and centralized command centers. Moreover, these digital tools are changing the way we perform traditional security operations as real time monitoring, predictive analytics and automated responses in the security operations realm enhance our preparedness in cases of emergency, uplift our crime rates and also earn us public trust in governance system (Kitchin,2016).

Among various technological improvisations in this sector, one of the main significant innovations is AI-powered surveillance which is based on the implementation of ML algorithms to analyze the video footage from the CCTV cameras in real time. Suspicious behavior can be detected, license plates can be recognized, unauthorized access can be identified and authorities notified automatically in these systems. London's "Ring of Steel" expansion of surveillance network armed with AI powered cameras, has been instrumental in counter terrorism and preventing crime since the 2010's (Clarke, 2019). Also, Chicago's "Strategic Decision Support Centers" (SDSCs) linking AI-driven data analysis with real time camera and sensors data allows police to see where high risk areas are and to deploy resources appropriately (Ratcliffe et al., 2019).

Another transformative feature that secures public spaces, transportation hubs, as well as high risk zones is facial recognition technology. Whilst this technology is controversial for privacy implications, it has been demonstrated by the identity resolution to identify suspects as well as locate missing persons, and to prevent threats. This is an example of how the Delhi Police used a facial recognition system in 2019 during a religious gathering in New Delhi and were able to identify nearly 3,000 missing children in four days (The Hindu, 2019). For instance, cities such as Shenzhen and Hangzhou combine facial recognition with AI to detect jaywalking, chase criminals, ensuring pedestrian crossings, etc. (Creemers, 2018).

Central to such smart city safety frameworks are integrated command and control centers (ICCCs). At these centers, an emergency services, traffic management, weather systems, surveillance networks and social media data is combined to form a unified situational awareness platform. Under Smart Cities Mission, Pune's ICCC was set up by the ICCC to connect city authorities to manage law enforcement, disaster response and utility services from a single dashboard (Smart Cities Council India, 2020). This platform came up as an important tool for monitoring compliance during quarantine, contact tracing, and even coordinating health services during the era of the COVID 119 pandemic.

Predictive policing, deemed a practice of using AI to analyze past crime data to predict future incidents, has made many cities adopt. The now discontinued "PredPol" software in Los Angeles uses the area's crime hotspots to predict and adjust patrol routes accordingly. While it was thought to have ethically problematic implications regarding bias and transparency, it showed an increasing role that algorithmic intelligence plays in urban safety planning (Brantingham et al., 2018).

These technologies have great benefits, but as with other systems of this nature will need to be carefully deployed in the face of strict data protection, privacy laws and public oversight to be used in an ethical manner. Smart surveillance reduces safety risks but is risky and over surveillance and loss of civil liberties is a threat to participatory governance and regulatory frameworks (van Zoonen, 2016).

Therefore, the AI enabled public safety systems help smart cities to become more resilient, secure and responsive. These innovations provide stronger public trust and optimize safety that allow a safer urban experience via prevention of threats in real time and improving emergency management, as well as supporting evidence-based policing. Responsible implementation, transparency, community involvement however all of their success will depend on.

## Sustainable Waste and Water Management

Waste and water management are important parts to a sustainable development of urban areas, and smart cities are currently turning to Internet of Things (IoT) technologies to improve these services. The use of IoT in waste and water systems enables real time data collection, predictive analytics, and automated control by city administrators for reduction of costs, optimal service delivery and higher environmental standard (Alabi & Bakare, 2019). With population living in cities already growing, these digital tools are so valuable on top of that.

In waste management and their smart bins are IoT enabled that comes with sensors to sense the fill levels, temperature, and location. Real time data flowing these bins is sent to centralized dashboards and waste collection services use this data to optimize route, cut fuel consumption and reduce those over flowing incidents. Not only does this increase operational efficiency, but it also improves urban cleanliness as well as greenhouse gas emissions for unwanted collection trips (Longo & Vetrano, 2021). Leading the way is Songdo, South Korea, where it brings its waste to processing centers via fully automated waste collection system inclosed underground pneumatic tubes from home without any need for traditional garbage trucks (Kim et al, 2019). This system has resulted in vastly reduced traffic congestion and environmental pollutions (related to waste collection).

Like Barcelona, smart bins are also installed in the city which reflects with the city's waste management system and so pickups are scheduled to happen only when they are needed. Bakıcı et al. (2013) state that the program, as part of broader smart city strategy, had resulted in more than 25% reduction in collection costs and increase in recycling rates. According to city data, Philadelphia's BigBelly solar powered trash compactors use sensors that signal when they are full and thereby create less frequent pickups by 75 percent in the U.S.

On the water side of the IoT sensors are being installed to detect water flow, leakage, water quality, and water distribution system regulation in the field. One of the more important applications of these technologies is in reduction of water lost through pipeline leaks, which is a major issue in many old urban infrastructure. Leaks are responsible for about 25 percent to 50 percent of water that does not reach consumers in urban areas, according to the World Bank (2016).

Motorola is helping Singaporean Public Utilities Board monitor the delivery of water through its use of IoT sensors and AI analytics. Abnormalities such as unexpected drops of pressure and that can indicate leaks, are detected by the system and perform maintenance before big damage come, as well. This led to Singapore's low water loss rate being around 5% (PUB Singapore, 2020).

In India, Pune and Hyderabad are among the early Indian cities to implement smart water meters and telemetry systems that allow households to track their use in real time and thus practice water conservation at the home level. They also help municipal bodies enforce equitable water distribution and are useful for detecting the illegal connections and tampering.

These smart systems not only raise the efficiency and conservation bar giving us better performance and cost savings but also have an impact on ensuring the higher goals of sustainability with less energy use, less operational costs and less environmental degradation. Incorporating of IoT into the urban waste and water systems is a key element to building resilient, adaptive and sustainable cities that can survive in population expansion and climate challenges as pointed out by Zanella et al. (2014).

## CHALLENGES IN SMART CITY IMPLEMENTATION

Despite huge potential of the smart cities to transform urban living through technological innovation and data led governance, but a few of the critical challenges facing smart cities still persist in making effective and long term implementation success. One of the most pressing issues is the cybersecurity vulnerabilities, data privacy concerns, and the task of integration of heterogeneous technologies out of different sectors in various jurisdictions (Kitchin, 2016; Neirotti et al., 2014). Challenges such as these have to be tackled in a multi-pronged, multi-stakeholder and multi-dimensional effort. This includes strong legal foundations, safe digital infrastructure, and unreserved cooperation among all the relevant stakeholders.

But when it comes to smart cities, identifying the most critical threat is cybersecurity. Today, there is a rapid rise of cyberattacks on urban infrastructure, which is getting increasingly reliant on digital technologies and interdependent systems: Smart grids, surveillance networks, public transport systems, etc. For example, the Atlanta, Georgia, city in 2018 was subjected to a major ransomware attack that brought down the city's funding operations including courts, as well as damaged through millions of dollars. (Satariano, 2018). Such incidents make smart systems vulnerable and thus require continuous monitoring, encryption protocols and incident response mechanisms.

Related to cyber security is the question of data privacy. IoT devices and sensors, mobile apps, surveillance tools, and other sources of smart cities produce enormous amounts of personal and behavioral data. This data is exposed or misused without proper safeguards, breaching the rights of the citizens. In the same year, San Francisco became the first city in the U.S. to ban facial recognition technology after public backlash against its deployment in cities like San Francisco (Singer, 2019). In the midst of growing fear of surveillance, and the desire to keep an eye on innovation versus civil liberties, this is exactly what's going on.

There is yet another great obstacle, which is the integration of various technologies. There are a wide range of digital tools in a smart city ecosystem produced by different vendors with different standards and protocols needed. It may be hampering lack of interoperability between systems, which may stop the systems from communicating seamlessly, and scaling. According to Townsend (2013), smart cities should use open standards and flexible architectures to enable the innovation and growth not dependent on vendor.

However, to solve these problems cities will have to develop and enforce strong legal and regulatory framework on how to collect and share of data. GDPR by the European Union is one of the examples and contains a comprehensive model for data protection which smart cities can adapt to local context (Voigt and Von dem Bussche 2017).

Similarly, it is equally important to develop a secured digital infrastructure such as ensured secure storage of data in the cloud, secured communication data networks and resilient communication systems. For instance, the implementation of blockchain is used for the transaction of secure data at the municipality of Barcelona's Urban Platform, which aims to increase both transparency and security (Bakıcı et al., 2013).

Lastly, the success of smart cities hinges on collaborative governance. Inclusive planning and execution in private enterprises, tech developers, and civil society organisations require governments to coordinate discreetly with them. Public private partnership has played a vital role in implementing energy, e mobility, and e governance projects with active community involvement through the pilot projects in the Smart City initiative in Amsterdam (van Winden et al. 2016).

Nevertheless, it is also clear that smart cities offer great opportunities for transformation, but also that cybersecurity, privacy, integration and governance issues must be addressed so that smart cities can be responsibly and sustainably developed. The barriers of smart urbanism such as this could be opened up, if with the help of strategic planning, policy making made inclusive and robust safeguards, cities could navigate these obstacles.

## CONCLUSION

While being a technological innovation, smart cities are nothing less than paradigm shift in the programme of urban planning, management and governance. Early conceptions of smart cities were anchored to connectivity and automations, however, today's understanding is that smart cities are cybermetabolism (or in technical terms that is, co-evolving systems of technology, people and process) where the complex urbanization issues to be tackled are addressed. The shift from siloed, reactive service delivery to one that is integrated, proactive, and citizen centered, is an indication of a transition to the kind of urban environments that constitute the urban age (Kitchin, 2016).

What makes a smart city is the fact that at its core, smart cities comprises of Internet of Things (IoT) sensors, Artificial Intelligence (AI), Big data analytics and cloud computing for optimization of operations and quality of life. However, they have the potential to be real transformers for sustainability, inclusivity, and resilience, which are the three driving forces of tomorrow's urban living (Batty et al., 2012). Smart cities technology is not a purpose in itself but a way to create evidence based decisions, increase community empowerment and environmental safeguard.

Amsterdam Smart City is, for example, a collaborative platform since 2009 which aims to merge government with private companies and citizens in co developing innovative urban solutions. It has started more than 80 pilot projects in energy efficiency, smart mobility and the area of digital governance. The project of one such "Climate Street" in Utrechtsestraat involved retrofitting buildings with smart meters and eco friendly technology, helping to decrease emissions, and engaging all local businesses in sustainable practice (van Winden et al. 2016).

Like in Songdo International Business District in South Korea, a purpose built smart city is also found. Songdo was constructed from scratch on reclaimed land from an area that was previously swampland and integrated ubiquitous sensor systems, pneumatic waste collection, centralized energy control, and the telepresence services in education and health care. The city has earned praises for its technological prowess, but likewise the critics have noted that the development of the infrastructure should be matched with the development of the community and the inclusion (Kim et al., 2019).

It includes smart city planning that facilitates technological benefits to everyone irrespective of the socioeconomic status. Integrated into the urban integration of far away protagonists like India's Smart Cities Mission, projects that upgrade slums with digital amenities as well as ensure equitable access to water, sanitation and public services may close urban divides. However, due to displacement and digital exclusion, concerns continue (Datta, 2018), and participatory urban governance which include voiced of marginalized people is needed.

The other main feature of smart cities is resilience, the capacity of the cities to respond to shocks and stresses, in order to continue to function. Cities need to be neither connected nor living alone if they are not adaptable. The sobering reality that climate change, pandemics and cyber threats have necessitated the ability to react to changing circumstances is real. UN Habitat (2020) explains that the City of Copenhagen has started to use a smart resilience approach of real-time flood monitoring as well as AI energy optimization to combat sea level rise and extreme weather. In response to the COVID-19 pandemic, Pune's India's Integrated Command and Control Centre (ICCC) transformed its traffic monitoring infrastructure into hotspots for the virus and facilities for medical coordination (Smart Cities Council India, 2021).

In addition, smart cities promote the sustainable environment by lowering emissions, energy conservation and efficient resource management. According to for instance Barcelona's smart lighting system the pedestrian activity determines whether the brightness decreases or is on the highest if there is no human activity. Singapore's Smart Water Grid is built on an integrated network of sensors that gathers information on leaks to detect and optimize water use, and contributes towards its aim of water loss of 5%, one of a few lowest rates in the world (PUB Singapore, 2020).

Although, it presents its challenges. Issues of data privacy, cybersecurity, interoperability, and technological dependency raise ethical and operational concerns. In fact, Townsend (2013) and van Zoonen (2016) agree that the smart city is not fully baked, unless it is technologically savvy, but also has transparent governance, solidified regulatory framework, and public engagement. True smart city transformation will occur only if smart cities are democratic, not technocratic, that is, the interests of citizens and not efficient system operation are being served.

It is also critical to smart city success that urban infrastructure evolves. The traditional infrastructure has to be digitally retooled or replaced entirely with sensor enabled, data based systems that can accommodate the emerging technology of autonomous vehicles, the decentralised power grids or monitor the real time environmental changes. It demands major investment, as well as capacity building of institutions interested in this, and alignment with the sustainability targets - including those of the UN Sustainable Development Goals (SDGs).

Things will continue to look ahead to another smart solution to help cities become smart, sustainable, and sustainable. The wider the cities adopt this transformation, the more nuanced and specific to context will be. The smart city is not a one size fits all, each urban area needs to craft based on its cultural, economic and environmental condition.

Smart cities, then, are an audacious view of urban life in the future, one where the best of human life will be supported, the ecosystem will be balanced, and the communities will be strong. To achieve this vision, however, it does not suffice with digital innovation; rather, the governance, equity, sustainability at city scale has to be reimagined. In the evolving cities, those who understand, practice and empower, this holistic, people centered approach will be in the best position to prosper in an increasingly complex world.

## REFERENCES


Alabi, O., & Bakare, A. (2019). Application of IoT in Smart Waste Management Systems. *International Journal of Environmental Science and Technology*, *16*(5), 2541–2552.

Albino, V., Berardi, U., & Dangelico, R. M. (2015). Smart cities: Definitions, dimensions, performance, and initiatives. *Journal of Urban Technology*, *22*(1), 3–21. DOI: 10.1080/10630732.2014.942092

Alyami, S. H., & Rezgui, Y. (2012). Sustainable building assessment tool development approach. *Sustainable Cities and Society*, *5*, 52–62. DOI: 10.1016/j.scs.2012.05.004

Ammad, U., Ahmad, A., & Qamar, F. (2020). Autonomous public transportation in smart cities: A case study of RoboBusLine in Helsinki. *International Journal of Transportation Science and Technology*, *9*(4), 283–294.

Bakıcı, T., Almirall, E., & Wareham, J. (2013). A smart city initiative: The case of Barcelona. *Journal of the Knowledge Economy*, *4*(2), 135–148. DOI: 10.1007/s13132-012-0084-9

Batty, M., Axhausen, K. W., Giannotti, F., Pozdnoukhov, A., Bazzani, A., Wachowicz, M., Ouzounis, G., & Portugali, Y. (2012). Smart cities of the future. *The European Physical Journal. Special Topics*, *214*(1), 481–518. DOI: 10.1140/epjst/e2012-01703-3

Brantingham, P. J., Valasik, M., & Mohler, G. (2018). Does predictive policing lead to biased arrests? Results from a randomized controlled trial. *Statistics and Public Policy (Philadelphia, Pa.)*, *5*(1), 1–6. DOI: 10.1080/2330443X.2018.1438940

City of Helsinki. (2020). *Helsinki 3D+ Digital Twin Project*. https://www.hel.fi

Clarke, R. (2019). Surveillance in the UK public sector. *Computer Law & Security Review*, *35*(4), 105332.

Creemers, R. (2018). China's social credit system: An evolving practice of control. SSRN *Electronic Journal*. DOI: 10.2139/ssrn.3175792

Crossrail Ltd. (2020). *Crossrail Annual Report*. https://www.crossrail.co.uk

Datta, A. (2018). The digital turn in postcolonial urbanism: Smart citizenship in the making of India's 100 smart cities. *Transactions of the Institute of British Geographers*, *43*(3), 405–419. DOI: 10.1111/tran.12225

Eastman, C., Teicholz, P., Sacks, R., & Liston, K. (2011). *BIM Handbook: A Guide to Building Information Modeling*. Wiley.

Fang, X., Misra, S., Xue, G., & Yang, D. (2012). Smart grid—The new and improved power grid: A survey. *IEEE Communications Surveys and Tutorials, 14*(4), 944–980. DOI: 10.1109/SURV.2011.101911.00087

FHWA. (2020). Adaptive Signal Control Technologies. *U.S. Federal Highway Administration*. https://www.fhwa.dot.gov

Guerra, E., & Morris, E. A. (2018). Cities, transportation, and the future of work: Insights from urban science. *Transport Policy*, *69*, 149–157.

Gungor, V. C., Sahin, D., Kocak, T., Ergut, S., Buccella, C., Cecati, C., & Hancke, G. P. (2011). Smart grid technologies: Communication technologies and standards. *IEEE Transactions on Industrial Informatics*, *7*(4), 529–539. DOI: 10.1109/TII.2011.2166794

Hashem, I. A. T., Chang, V., Anuar, N. B., Adewole, K., Yaqoob, I., Gani, A., Ahmed, E., & Chiroma, H. (2016). The role of big data in smart city. *International Journal of Information Management*, *36*(5), 748–758. DOI: 10.1016/j.ijinfomgt.2016.05.002

IEA. (2020). *Smart Grids: Tracking Report*. International Energy Agency. https://www.iea.org

IEA. (2023). *Global EV Outlook 2023*. International Energy Agency. https://www.iea.org/reports/global-ev-outlook-2023

Kim, M., Lee, J., & Lee, H. (2019). Smart cities and urban policy in South Korea. *Cities (London, England)*, *89*, 48–56.

Kim, M., Lee, J., & Lee, H. (2019). Sustainable urban infrastructure through smart waste management. *Journal of Cleaner Production*, *220*, 123–134.

Kitchin, R. (2016). The ethics of smart cities and urban science. *Philosophical Transactions. Series A, Mathematical, Physical, and Engineering Sciences*, *374*(2083), 20160115. DOI: 10.1098/rsta.2016.0115 PMID: 28336794

Leape, J. (2006). The London congestion charge. *The Journal of Economic Perspectives*, *20*(4), 157–176. DOI: 10.1257/jep.20.4.157

Lee, J., Park, J., & Kim, H. (2018). Structural health monitoring for urban bridge safety in Seoul using sensor technologies. *Sensors (Basel)*, *18*(3), 845. PMID: 29533994

Longo, F., & Vetrano, A. (2021). Smart Waste Management in Smart Cities: A Decision Support Approach. *Simulation Modelling Practice and Theory*, *108*, 102275.

Ministry of Housing and Urban Affairs, Government of India. (2022). *Smart Cities Mission Annual Report*. Retrieved from https://smartcities.gov.in

Neirotti, P., De Marco, A., Cagliano, A. C., Mangano, G., & Scorrano, F. (2014). Current trends in Smart City initiatives: Some stylised facts. *Cities (London, England)*, *38*, 25–36. DOI: 10.1016/j.cities.2013.12.010

Pecan Street Inc. (2019). *Data-Driven Energy Solutions*. https://www.pecanstreet.org

Pierce, G., & Shoup, D. (2013). Getting the prices right: An evaluation of pricing parking by demand in San Francisco. *Journal of the American Planning Association*, *79*(1), 67–81. DOI: 10.1080/01944363.2013.787307

Ratcliffe, J. H., Strang, S. J., Taylor, R. B., & Sorg, E. T. (2019). The impact of intelligence-led policing in a major U.S. city. *Justice Quarterly*, *36*(5), 831–856.

Satariano, A. (2018). Atlanta's Cyberattack and the Age of Digital Disaster. *The New York Times*. https://www.nytimes.com

Singapore, P. U. B. (2020). *Smart Water Management*. https://www.pub.gov.sg

Singer, N. (2019). San Francisco bans facial recognition technology. *The New York Times*. https://www.nytimes.com

Smart Cities Council India. (2020). *Integrated Command and Control Centres: Case Studies*. https://smartcitiescouncil-india.com

Smart Cities Council India. (2021). *Smart City Initiatives and COVID-19 Response*. https://smartcitiescouncil-india.com

Smart Nation Singapore. (2020). *Smart Mobility Initiatives*. https://www.smartnation.gov.sg

Smart Nation Singapore. (2020). Smart Nation Initiatives. Retrieved from https://www.smartnation.gov.sg

Smart Nation Singapore. (2020). *Virtual Singapore Initiative*. https://www.smartnation.gov.sg

Teicholz, P. (2013). BIM for facility managers. *Journal of Building Performance*, *4*(1), 17–28.

The Hindu. (2019). Delhi police use facial recognition to identify missing children. https://www.thehindu.com

Townsend, A. M. (2013). *Smart Cities: Big Data, Civic Hackers, and the Quest for a New Utopia*. W. W. Norton & Company.

UN-Habitat. (2020). *Resilient Cities and Climate Action*. https://unhabitat.org

United Nations. (2018). World Urbanization Prospects: The 2018 Revision. Retrieved from https://population.un.org/wup/

Van Winden, W., van den Buuse, D., & van Dijck, E. (2016). Smart city pilot projects: Exploring the challenges and lessons learned. *Journal of Urban Technology*, *23*(4), 3–20.

Van Zoonen, L. (2016). Privacy concerns in smart cities. *Government Information Quarterly*, *33*(3), 472–480. DOI: 10.1016/j.giq.2016.06.004

Voigt, P., & Von dem Bussche, A. (2017). *The EU General Data Protection Regulation (GDPR): A Practical Guide*. Springer. DOI: 10.1007/978-3-319-57959-7

World Bank. (2016). *Reducing Non-Revenue Water*. https://www.worldbank.org

Zanella, A., Bui, N., Castellani, A., Vangelista, L., & Zorzi, M. (2014). Internet of Things for Smart Cities. *IEEE Internet of Things Journal*, *1*(1), 22–32. DOI: 10.1109/JIOT.2014.2306328

# Chapter 9
# Edge Computing for Energy-Efficient Internet of Things: Concepts, Technologies, and Applications

**Delsi Robinsha S**
https://orcid.org/0009-0002-8535-5662
*Department of Computing Technologies, SRM Institute of Science and Technology, India*

**B. Amutha**
https://orcid.org/0000-0001-9875-4852
*Department of Computing Technologies, SRM Institute of Science and Technology, India*

## ABSTRACT

*In today's world, the utilization ratio of IoT devices has been increased in the production of massive data, and these gigantic problems were posed in energy consumption and environmental conservation. To address with these problems, there is an innovative concept known as the edge computing that means computation is carried closer to the place where the data is collected. This is done in view of the above stated challenges and difficulties. Building the concept of 'Efficient Edge For Sustainable IoT' As a response to the presented research questions, this chapter employs the energy-efficient edge computing architecture concept to explore the Efficient Edge as a sustainability strategy for the IoT. It elaborates crucial questions related to edge computing and reveals the position of this concept for minimizing the energy consumption, enhancing resource efficiency, as well as the rational data*

DOI: 10.4018/979-8-3373-2802-7.ch009

*processing on the end side of the network. Some of the diverse technologies and strategies expounded in this chapter to achieve energy efficiency in edge computing include energy-aware task offloading, load balancing as well as the integration of renewable energy. Besides this, it looks at the joy of optimization and solutions on battery-powered Internet of Things devices and issues on scalability for edge computing. In this chapter, the focus is made on how the Efficient Edge allows IoT ecosystems to be sustainable, scalable, and resilient, while the use of real-world case studies and the analysis of the future looks like.*

## 1.1 INTRODUCTION

Recently, IoT has introduced change in the way how people communicate with environment. This has created a huge array of networked devices that can in any given moment collect, exchange, and analyse data. However, this exponential increase in the use of Internet of Things has created massive problems especially in handling data, connection as well as power consumption. Regarding the data storage and processing what catches the researchers' attention is that the traditional IoT architectures often employ cloud servers. This type of architecture can lead to latency issues, bandwidth constraints and raised power consumption related with the exchange of data through networks. The essence of edge computing has emerged as the revolutionary way of managing decentralized data processing and computation to address the problems that have been enumerated (Premalatha & Prakasam, 2023). Edge computing is therefore a subcategory of computing that entails the processing of computations and data at the network edge where data is commonly produced. This can therefore cut down on the delays, increase the speed of service delivery and there by efficiency is enhanced. Currently, it is conceivable for Internet of Things (IoT) systems to achieve significant enhancements in the areas of scalability, dependability, and power proficiency through edge computing. It will, therefore, create grounds for making the Internet of Things environment more sustainable and reliable.

In this chapter, a comprehensive analysis of how edge computing and IoT are related is described. The main emphasis of this research is placed on such framework as key concepts, structures, techniques, and solutions stimulating this shift. In this article, we induce the main causes that have made the use of edge computing in the Internet of Things adoption. These considerations include processing needs to be done concurrently, higher number of devices requiring connection and notification, and a higher need for application that has less delay and needs high bandwidth (Liu et al., 2024).

This Chapter focuses on edge devices, edge servers, and fog computing for distributed computing and big data analysis at the edge of a network. Also, it focuses on understanding the constituent parts and operation of edge computing architectures. Further, it examines the challenges and prospects of putting into practice edge computing in today's Io Things structures. It is most concerned with scalability and interoperability as well as security due to its strategic importance to adoption and innovation.

In this chapter, these use cases and examples of real-world applications enhance and explain how edge computing has revolutionised several industrial sectors. These verticals are smart cities, health care, manufacturing, transportation and the list can go on and on. However, the main objective of this work is to provide the reader with a holistic view of how this convergence will redefine the future of connected systems for the enhancement of efficiency, flexibility, and sustainability in the digital age. This will be done through the analysis of interactions between edge computing and the Internet of Things.

## 1.2 ARCHITECTURES AND FRAMEWORKS FOR EDGE COMPUTING

Due to the nature of handling data closer to the point where it is generated, edge computing architectures and frameworks have been realized. They have been evoked in a bid to minimize latency, minimize bandwidth, and maximize real time scenarios. For enhancing the energy efficiency for the IoT systems, edge computing architectures and frameworks are special significant. Edge computing reduces the amount of communication required with the central cloud servers; this serves the purpose of conserving bandwidth and minimizing the energy consumption. This is done through a technique that brings processing closer to the data sources.

### 1.2.1 Edge Computing Architecture Components

Computing designs of the network decentralise equipment placement shifting them from a single company data centre to multiple edge sites. This is the principle on which most edge computing designs are based upon or rely on. This encompasses storage and application and sensors, and computing assets (Saadallah & Alabady, 2024). All of these are located in the periphery and are linked with a central data centre or cloud over a network. On this periphery, information is collected, analysed or both through the use of gadgets and probes. As beneficial equipment's these devices can collect data, process, and execute it in real time that is sufficient enough to overcome the limit of network help in most of the cases. Simplistically constructible yet at a comparatively tiny distant location is a reduced form of on-premises edge server or data centre that can be established. This produces topology possibilities that can still accommodate many different environmental conditions and at the same time cut footprints.

*Figure 1. Architectures and Frameworks for Edge Computing*

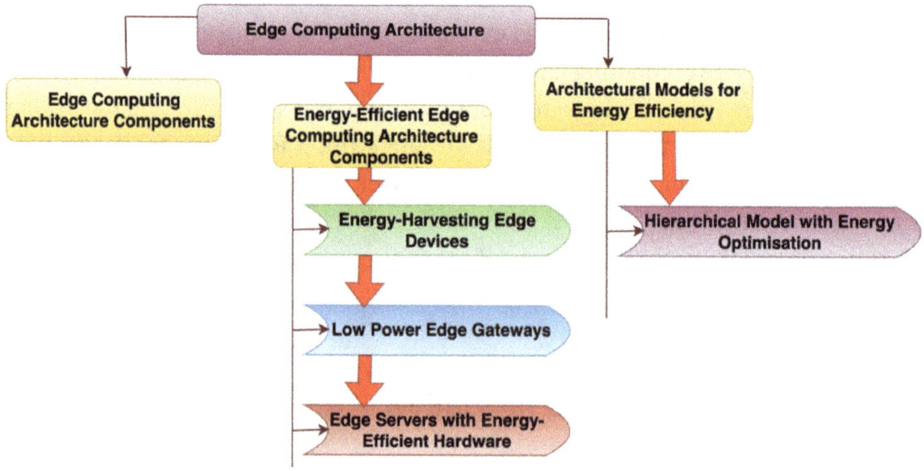


## 1.2.2 Energy-Efficient Edge Computing Architecture Components

Energy-Harvesting Edge Devices:

The technologies that are used in these gadgets include; Solar cells, Piezoelectric and Thermoelectric converters that produce electricity from the surrounding environment.

Low-Power Edge Gateways:

These gateways help in the easy management of data aggregation, initial data processing, as well as interfacing with the edge devices and cloud servers while being designed to consume the least amount of power (Chakraborty et al., 2023).

Edge Servers with Energy-Efficient Hardware:

Lowering of the local edge server energy utilization by employing low-power central processing units, power-conserving storage devices, and improved cooling mechanism.

### 1.2.3 Architectural Models for Energy Efficiency

Various edge computing architectural models are designed in a way that aims at consuming less power and at the same time ensure that the gathered information is processed in a reliable and efficient manner. As for the identified models, to ensure sustainability, control power consumption, and achieve high performance in IoT applications, the potential methodologies are used.

#### Hierarchical Model with Energy Optimization

The hierarchical paradigm is used to classify computing resources by several levels owing to the fact that each layer is charge with a different role of processing of data and their administration.

- Edge Devices, Also Known as Tier 1 Sensors, actuators, and Internet of Things devices are the principal sources of data available. Some of these energy efficient features are mainly like Power management, the duty cycle that is used that requires the devices to be turned on and off at certain specified intervals. Other related topics are low-power microprocessors, energy scavenging, and power control methods.
- The second tier is the Edge gateways. Intermediate devices are involved in data collection before the data is transferred to the local edge servers or the cloud, information acquisition and initial processing as well as filtering. Compared to each node of the distributed system, it is endowed with larger capability of store and compute (Milani et al., 2024). Therefore, to reduce energy consumption it is required to implement adaptive sampling, which means that the rates of data collection vary according to the demand of the case at hand.
- The third tier is the local edge servers. Since it performs large processing workloads as well as stores data, it is more towards the edge than actual cloud data centres. In a bid to reduce energy consumption and arsenal the gear is in a position to consume, utilize low energy processors and cooling systems that have been toned down.
- Techniques include energy and duty cycling which imply powering and depowering of these devices in accordance to their usage. In adaptive sampling the rates at which data is generated are varied depending with the situation and the circumstances in the environment. This maximises the amount of energy that is used without a corresponding reduction in the quality of the information eaten. When it comes to energy-aware task scheduling, the tasks

are scheduled with great regard on the occurrence of energy and consumption profiles thus enhancing efficient use of resources.

## 1.3 TECHNOLOGIES ENABLING EDGE COMPUTING

Edge computing is a distributed computing architecture that avoids the high-central data centre or the cloud by employing data processing and storage nearer to the place where the data were initially generated. Cutting down the bandwidth usage, increasing responsiveness, and permitting the timely data analysis and response generation are the objectives of this approach. To make edge computing a reality, a number of essential technologies and components are required:

### 1.3.1 Edge Devices and Hardware

In edge computing, a variety of specialised hardware devices placed at the network's periphery serve as its backbone. Among these peripherals are:

#### Internet of Things (IoT) Sensors and Devices

This includes a diverse mixture of so-called Things or devices at the edge of the network that generate information and are connected to the Internet such as cameras, industrial machinery, or sensors. Depending on how you define the gadgets, these can range from simple sensors to complex machines or systems (Alsharif et al., 2024).

#### Edge Gateways

Gateways at the network's periphery act as intermediaries, collecting data from a variety of Internet of Things (IoT) sensors and storing it locally before sending it to the cloud. Gateways in the network's periphery provide secure connection, device control, and data preparation before it reaches the cloud.

#### Micro Data Centres

More computing and storage capacity than individual edge devices, housed in small-scale, modular data centres closer to end users. These smaller data centres connect the edge with the cloud and can process more complicated data.

### Mobile Devices

A wide variety of portable computing devices, including smartphones, tablets, and e-readers, are capable of carrying out edge computing operations (Robinsha & Amutha, 2024). Because mobile devices can do data processing locally, they don't always need to be online.

### Embedded Systems

To enable processing and control at the edge in real-time, embedded systems integrate specialised computers into bigger devices or machines. Autonomy and the ability to function independently of external resources are hallmarks of embedded systems, which are typically developed for particular purposes.

## 1.3.2 Communication Protocols

Probably one of the key parameters of edge computing is the data transfer rate from devices located at the edge of the network to the cloud or central computing centres. Cellular 5G and LTE networks and Wi-Fi 6 and the protocols MQTT and OPC UA instruments are indispensable for edge computing to work. For connecting the various mobile edge devices in real time, nothing could be better suited than 5G and LTE as they ensure lower latency rate and high speed to meet the needed bandwidth in the edge. Thus, for enhancing the reliability and efficiency of the edge networks, especially in the crowded conditions, the next generation wireless standard, Wi-Fi 6, is known to support improved low-latency throughputs for the local area edge network. Intended for use over unreliable connections and low bandwidth, MQTT is eminently suitable for use with edge computing because it is optimized for IoT and edge uses. Used for edge automation and control, OPC UA is a communication framework mainly built for the reliable and secure transmission of data from the edge to the cloud in real-time. Besides, the synergy effects of various communication technologies might enable the realization of edge computing's full potential by optimizing data exchange between devices at the edge and the cloud or other central data facilities.

## 1.3.3 Software and Middleware for Edge Computing

Decentralized edge computing systems require software and middleware platforms to manage, orchestrate, and secure the systems. The edge computation infrastructure mainly involves efficient real-time operating systems designed for continuous computing devices and edge gateways. Examples of such distributions are Yocto Project and Wind River Linux both of which are based on Linux operating system. These operating systems for the edge are developed with an eye on the basic computational capabilities of the edge devices while at the same time being able to provide all the required functionalities. Kubernetes and Docker Swarm are two examples of edge orchestration technologies that allow deploying, running and scaling applications on the edge environments (Sisi & Souri, 2021). These platforms also bring efficiency, making containerisation of edge applications as well as management straightforward and resource optimisation of the computing system as the demand fluctuates. Those are edge analytics solutions that are needed prior to data transmission to the cloud for processing, analysis and extracting insights from them. These technologies can concatenate, condition, and select data in an effort to enhance the system performance and reduce the amount of data sent to the cloud. Finally, the protection of the edge and its devices, data, and communications is necessary for countering cyber threats. Due to the shortcomings such as decentralized and often remote installations, it is imperative that these solutions include authentication, encryption, and anomaly detection to ensure the security of edge computing in the system. Exercising the full potential of this distributed computing paradigm the bundling of these various specialised software and middleware elements enable efficient control, coordination and security of edge computing environments (Khan et al., 2024) .

*Figure 2. Technologies Enabling Edge Computing*

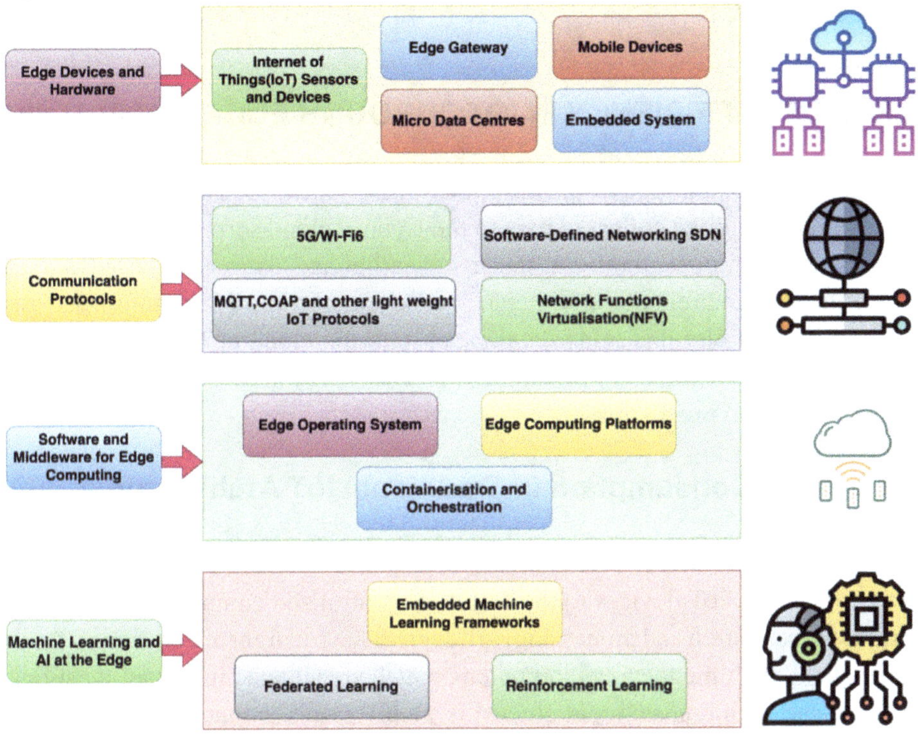


## 1.3.4 Machine Learning and AI at the Edge

Embedded machine learning makes edge computing possible since it allows a learned AI model to be used immediately on the edge device for making conclusions and decisions on data. Open-source frameworks like TensorFlow Lite and PyTorch Mobile enable and facilitate training and using of models on resource constraint edge devices. This paves way for low latency processing on the actual physical device to be made possible. Distributive learning being one of them, Federation learning is capable of training AI models across multiple edge devices without sharing raw data (Quy et al., 2023). This method allows different edge devices to collaboratively fine-tune a model on data that is kept local to such devices. This makes privacy better and reduces the number of data that has to be transferred to the cloud for storage and or processing. They are predestined to perform machine learning tasks at the edge by means of dedicated hardware accelerators called 'edge AI chips'. Some of them are NVIDIA Jetson which is development boards and Intel Movidius which provides special purpose ICs for deep learning inferencing at edge (Robinsha &

Amutha, 2023). This ultimately makes the AI processing on the device faster and less power-consuming.

## 1.4 ENERGY EFFICIENCY IN IOT THROUGH EDGE COMPUTING

The energy consumption in the traditional IoT architecture is large because the devices always send the collected data to the Central cloud servers for processing. This is quite energy consumptive as discussed earlier and also has the disadvantage of higher latency and reliance on a stable internet connection. For instance, sensors located in a smart city that relay data instantly to the cloud or health and fitness wearables located on humans would utilize a lot of energy because of the perpetuity of data transfer and processing.

### 1.4.1 Energy Consumption in Traditional IoT Architectures

In most conventional IoT designs, most power is consumed during the transmission of data by IoT devices to the clouds or centralized computing centers, not during data acquisition and processing. IoT devices have to transmit enormous flows of raw data all the time to cloud computers, which are situated in distant locations. A major issue with conventional cloud IoT architectures is that this technique of data transmission is quite power-intensive. The embedded computers typically have restricted power, storage, and battery capacity in IoT systems. This makes the high energy costs of sending data all the time as a big issue to be addressed. Calling the cloud and waiting for responses also causes some latency which can compromise IoT apps that require the apps to process information quickly. It is pointed out that due to power consumption, tendency to slow and generally high requirement toward standard IoT architectures, more efficient and revolutionary methods should be provided for solutions like edge computing, which can occur data processing and analysis near to its origin. They believe that edge computing holding back a significant part of the data processing to the local device can reduce energy consumption in the IoT systems.

### 1.4.2 Role of Edge Computing in Reducing Energy Use

Reducing the consumption of energy is part of the strategy that includes as one of the measures the implementation of Edge computing. The aspect of energy efficiency also come into play because of the fact that edge computing brings processing closer to where data is produced or processed. This means that data will not need to be sent all the way to distant cloud data centers as frequently. When data is analysed locally in edge devices, they send out less amount of data through the network, which definitely requires less energy. This is one of the main factors that use of edge computing reduces energy consumptions as mentioned above. It has often been said that cloud data centres consume lots of energy, in comparison, edge devices are designed to use less. They possess their unique hardware and software that allows them to work with less power, but process data. Additionally, similar to preventing data from having to go to the cloud and back, edge computing also allows decisions to be made near where the data is being generated. This is helpful in management since it reduces the energy that was required to transmit data to and from the center. In addition to that, edge computing platforms can alter how exactly available processing power is allocated for workload and can disable unneeded components to save power. In addition, the architecture of the edging computing can be powered by solar, or wind where available meaning the devices are built to be even less reliant on the power grid in general. Therefore, edge computation is one of the ways through which the application of technology can be made more environmentally sensitive.

### 1.4.3 Techniques for Energy-Efficient Data Processing

There are various ways through which energy efficiency in edge computing settings for processing data can be attained. Data preparation being one of the most critical strategies employed in intelligent edge systems, edge devices are used to filter, aggregate and even compress data, so that only a limited amount has to be transmitted to the cloud. This reduces the power that is needed for transmitting information to a certain level. There are other techniques that can be employed in the edge computing system such as adaptive resource orchestration that shuts down or degrades those resources in a system that are not heavily utilized in order to conserve energy depending on the amount of load or energy constraint placed on a system (Ji et al., 2022). Actually, the specific construction of hardware also requires very high energy efficiency. For instance, edge devices should contain low-power processors, low power consumption memory and its best-in-class power control mechanism in order to optimise its power usage. By applying these methods, edge computing systems can significantly reduce identified energy consumption and the pollution rates and still benefit from the spread data processing with low latency.

### 1.4.4 Case Studies on Energy Savings

#### Smart Grid Management

Having demand management and grid optimisation as the means to serve the smart grid, edge computing provides an important part of the path way. Making possible the supply and demand information of energy at the precise time to enable the supplier match the renewable energy generation with the consumers in the vicinity, edge-enabled devices give this a go. Edge computing allows microgrids to instantly measure the availability of restricted renewable resources including the solar and wind power meaning that microgrids can generate adequate amounts of renewable energy to fulfil the demand in a specific region.

#### Predictive Maintenance in Manufacturing

Due to edge computing, it is possible to enhance the concept of using prediction for maintenance requirements in manufacturing plants and other industrial settings. Distributed edge computing devices are able to detect the overall health status of the machinery in real time and predict when a particular machine needs to be serviced because they receive data from various sensors installed in equipment surrounding them. As a result, it reduces avoidable losses of working time and ensures that the equipment operates with the maximum possible energy output. The system is well equipped to act quickly when the problems are identified and amendments made to try and minimize the amount of energy loss from the system.

#### Autonomous Vehicles

Autonomous cars also use edge computing to make decisions in real time and management of energy efficient. Offload artificial intelligence procedures the data received by the car's onsite cameras and LIDARs and other gadgets thus making it possible for the car to respond to changes in driving conditions. As a result, it becomes possible to have the degree of efficiency of using energy at the level as high as possible; thus, it is possible to adjust the battery management system of the vehicle depending on the style of driving as well as the conditions of the road. Another advantage of edge computing is that platooning of trucks becomes feasible, and this could indeed mean that all trucks may not need a driver except, of course, the lead vehicle. This would be accompanied by massive savings on gasoline (Idrees & Jawad, 2023).

## 1.5 APPLICATIONS OF EDGE COMPUTING IN IOT

Edge computing in the context of IoT refers to the process of processing data by the devices close to the place where the data is generated rather than central cloud. Thus, this method reduces the latency time, increases the response time and makes the info more secure and personal. Real time monitoring and Predictive maintenance is done using Industrial IoT Smart homes and building using IoT for managing energy Health care analyzes health data as soon as possible Self driven cars assist people in taking quick decisions about their travelling Precision farming and environment monitoring is done using IoT in Agriculture. It makes IoT systems work better and more efficient through processing data a little closer to where it is generated (Khanh et al., 2023).

*Figure 3. Applications of Edge Computing in IoT*

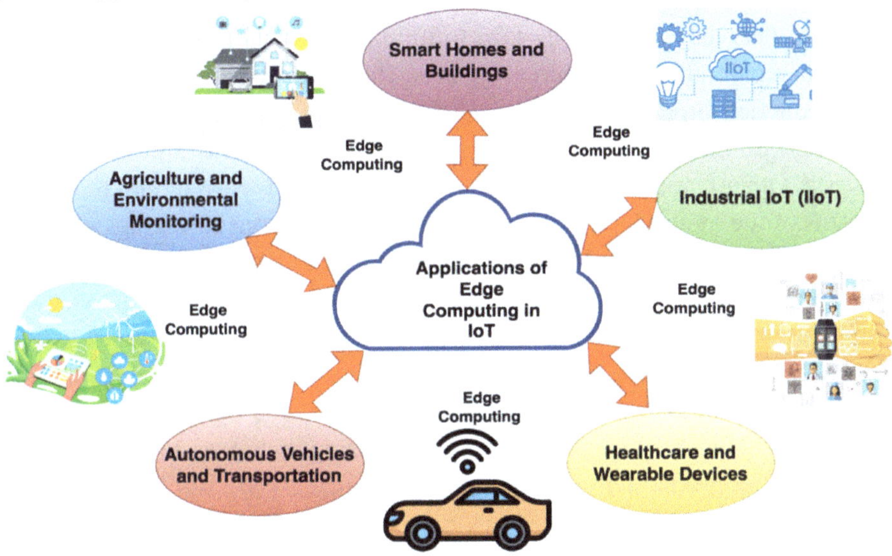


### 1.5.1 Smart Homes and Buildings

Smart home and building automation systems can only be effective with edge computing being at the core of them. They are able to control efficiently the energy consumption by adjusting automatically aspects like HVAC, lighting, and appliances based on present occupancy, climatic conditions, and usage profiles. This is made possible by the fact that these systems process the data on the edge devices merely.

Shorten the response time associated with this module so that the security systems can detect and deal with emergencies or an intruder, in a more efficient manner. To look for anomalies, predict the required maintenance time, and program the maintenance for various building equipments; the data from sensors should be processed (Vinothkumar et al., 2023). When it comes to smooth advanced user interfaces which require less time for processing, voice control, facial recognition and similar features should be activated. Smart homes/buildings will be able to be more reactive, less power-hungry, and also much safer with edge computing. It does this by enabling data and decisions to be made at the decentralized level.

### 1.5.2 Industrial IoT (IIoT)

By harnessing the use advanced Applications of the Industrial Internet of Things, Edge devices are poised to bring the capability to discover, predict and remedial problems with industrial equipment from the analysis of data from sensors within the device. This they specifically lead to minimization of time spent by a unit or machine in repair and in turn increasing efficiency. The opportunity to monitor and adapt production processes in real-time, provided by edge computing, results in the process optimization, which results in better control, quality and, consequently, less waste. Other features of edge gateways are to provide insight into the location and status of the industrial assets; it can integrate data from RFID tags along with sensors to inform on asset tracking. This enhances transportation and also usage or intake as it may be called depending with the needs of the business. Higher workplace safety and compliance could be attained by integrating smart wearable edge devices, these, will be able to monitor, our vital signs, surrounding conditions and equipment usage. If industrial organisations move intelligence to the edge, that means they will be able to make decisions faster and with better data, which in turn equates to better productivity, quality, and efficiency.

### 1.5.3 Healthcare and Wearable Devices

The advance in edge computation is transforming the healthcare sector through the incorporation of networked medical appliances and wearables The use of wearable edge appliances makes it possible to do distant patients' monitoring by frequently surveying their vital signs, activity, and prescription compliance. This creates early intercession; therefore, better results can be achieved. It is felt that edge computing conducts real-time video consultations and remote diagnostics hence low latency and privacy. This is made possible by handling data locally and this falls under the category of edge computing (Saleem et al., 2023) . Automated diagnostics means the ability of the edge devices to analyse medical images, data from laboratory tests, and etc. to help the doctors in diagnostics to be more accurate and faster. Support for Clinical Decisions: Machine learning that is implemented on the edge can provide a decision based on data from patient's records and the latest advancements in the medical field to the healthcare practitioners in real time. Edge computing can reduce issues surrounding data privacy since they are solved locally through processing of sensitive heath care data and enable the employment of applications that are critical to life at the edge of the network (Konecny et al., 2023).

### 1.5.4 Autonomous Vehicles and Transportation

The emergence of connected-big and small vehicles can only be due to the use of edge computing as a key enabler for think and self-driving cars Edge devices in vehicles can be used to interpret the real world via cameras-LIDAR-radar in the shortest time possible to make decisions on avoidance of collisions and safe spinning. Predictive Maintenance: The heightened capability of edge computing include ability to diagnose health of vehicles, and predict the right time to perform a maintenance check besides fuel efficiency which can lead to loss of operating expenses and potentially harm the environment. Systems installed on infrastructure such as traffic lights and other devices in vehicles have the possibilities of analyzing traffic flow and change the signals for improvement a the traffic flow. This is referred to as traffic management Differentiating between these two conceptions, one can construct a traffic management that advertises achieving certain goals and at the same time does not require these goals to actually be attained (D. Liu et al., 2022). Shipment may be tracked, route which the vehicle takes can be altered, and real-time information can be delivered to logistics companies with edge gateways that can be placed on vehicles and cargoes. For the reliable and efficient operation of the self-organized and interconnected transportation systems, the characteristics of the edge computing such as low latency, and localized processing are essential (Zhu et al., 2022).

### 1.5.5 Agriculture and Environmental Monitoring

Edge computing allows for vast opportunities in the IoT applications in agriculture and environmental protection; The right amount of water to apply and fertiliser to use, and other useful aggravations, taking into account information from soil sensors, weather stations, and drone images, can be governed by edge computing devices. This leads to precision farming In this case, farmers use modern techniques in soil management, crop varieties, irrigation, fertilizing, pest control, among others in practicing precision farming For instance, precision farming involves applying fertilisers based on the information retrieved from the database. HR monitors for animals can be placed on them in the form of wearable devices and the information gathered includes health status, location and behaviour to help farmers detect issues early enough and improve on the management of their herds. This is commonly referred to as livestock monitoring software (Bhatia, 2022). There is a way of following animal density, detecting cases of poaching and also give support to the cause of conservation through AI edge cameras and Remote sensors that are placed in hard to reach areas. Environmental Sensing: Edge nodes are capable of processing data from environmental sensors like air, water and other sensors for real time analysis for waste management and prevention of pollution. With the help of edge computing, it would enable the creation of IoT solutions that are more durable, energy efficient, and suitable for deployment in scenarios that are considered resource limited or in remote areas. This is done by performing computations independently and reduce the dependence on PC or internet connection.

## 1.6 SECURITY AND PRIVACY CONSIDERATIONS

Security and privacy is one area in which edge computing must consider, because there will always be sensitive data which must be secured when analysed at the edges of networks. Other elements of data security include components such as encryption, authentication, and secure boot processes that protect information from hackers and other people who must not view it. Privacy-related risks are addressed by minimizing the kinds of data collected and applying anonymization and pseudonymization where possible and ensuring compliance with data protection laws. To protect the data being sent from the edge devices to the central servers, there are reliable communication methods such as SSL/TLS, MQTT and CoAP are used (Li et al., 2022). It is essential to state that the effectiveness of these measures has been successfully illustrated by practice in various IoT frameworks, including smart homes, healthcare centres, and factories.

### 1.6.1 Data Security at the Edge

Ensuring the edge data that is generated, processed and stored on edge device is safeguarded from data breaches, cyber issues and unauthorized access is part of the steps taken for data security at the edge. One of the measures that can be employed ensures that data is not easily accessed by unauthorized persons include the following; Data has to be encrypted when stored as well as when being transmitted in the network. Authentication protocols across two or more factors and other stringent variations in the control of permissions do guarantee that only specifically approved parties can get through to edge resources. Firmware enhancements and required boot can be useful in protecting edge devices and preventing hackers from exploiting the devices. Also, the applications with intruder detection and prevention systems monitor for security threats and respond immediately to reduce the danger (Kong et al., 2022). For instance, within a smart factory encrypted communication and secure boot could be used to ensure that hackers cannot access private production data.

### 1.6.2 Privacy Concerns and Solutions

Minimal data techniques acquire only the requisite data for functional uses and process it at the place of acquisition. This can be useful to solve privacies on edge computing. The amount of personal data transferred to central computers is limited, which, in turn, ensures users' privacy. Data masking is a technique of obscuring the data before the time it reaches the edge and anonymization and pseudonymization augment the protection of people's names. Users' privacy is protected through informing them as to how their data is collected and obtaining their consent as well as allowing them to control their privacy options. It is also crucial to follow data security rules such as GDPR and CCPA to guard the users' privacy and avoid the law (Mahenge et al., 2022). For instance, in healthcare edge computing ensures that patients' data is processed locally so as to meet the necessary HIPAA requirements as well as to protect patients' privacy.

### 1.6.3 Secure Communication Protocols

To ensure that data transferred between and among edge devices, gateways, and central servers cannot be intercepted and read or modified, secure communication modes should be employed. IoT settings are effective using protocols including; MQTT and CoAP that actually provide safe ways or encryptions such as SSL/TLS for MQTT and DTLS for CoAP (Minh et al., 2022). Most people today employ HTTPS and SSL/TLS to secure the traffic being send or the information being transmitted whilst in the cyberspace (Mao et al., 2017b). Virtual private connections and other secure tunnels allow to ensure that the traffic going through public connections is secure. This of course makes it possible for edge devices to be accessed safely over the internet. There are two protocols frequently used in smart home technology which are Zigbee and Z-Wave. They also present secure methods of communicating since details must be protected (Wadhwa et al., 2022). MQTT with SSL/TLS encryption can be used to securely transmit data from edge sensors to the centralized SCADA in a smart city project (Xiang et al., 2021).

## 1.7 FUTURE DIRECTIONS AND TRENDS

The development of the Artificial Intelligence, Machine Learning along with the implementation of 5G network and the enhancement of traffic in real time data processing is a better picture towards the future of Edge computing inside the IOT. The decentralised methodologies will then be improved to minimize the computation in the cloud whereas AI at the edge will improve analysis and the authority to make decisions on the gadgets. New generation Internet of Things (IoT) applications will get even more efficient and faster with help of the main 5G networks driver increasing bandwidth and cutting the latency and improving the connectivity for Edge computing even more. Another area of concern is the proper interaction where a protocol has to be defined for making different things in the Internet of Things work together; and another component is the incorporation of the Blockchain to ensure that the transactions are secure and the data is not tempered with. The technologies like AS, AR, and VR will revolutionise several domains such as smart cities, healthcare, transportation, and manufacturing, etc Edge computing will be one of the significant components of these upcoming technologies. Analysing all these trends, it is needs to be pointed out that edge computing remains and will remain one of the key factors that define IoT solution scalability.

## 1.8 CONCLUSION

Edge computing is an revolutionary method of broadening the functionality of IoT systems. How it functions is through operation in layers that are closer to the source of data and this fosters the reduction of latency and also enhances the level of performance of the system. In this study, we have seen how the edge computing applies challenges like security and privacy, as well as the need for real time decision in various applications like smart homes, industrial IoT, healthcare automation, self-driving automobiles and agriculture. Advancement of more effective and reliable IoT solutions owe to machine learning, AI, 5G and standard protocols which is good for edge computing. More innovation in edge AI, distributed computing, and other applications extending to smart cities, environmental monitoring, and others are expected with other industries adopting edge computing solutions. More accuracy and less cash outlay are just the tip of the iceberg, and the future recipes of Edge Computing will let people and companies unleash the application of IoT in an interconnectivity of a smarter world.

**Author Contributions** Conceptualization, Delsi Robinsha S.; Methodology, Delsi Robinsha S and B.Amutha; Software, Delsi Robinsha S; validation, B.Amutha; formal analysis, Delsi Robinsha S.; investigation, B.Amutha.; resources, B.Amutha.;

data curation, Delsi Robinsha S.; writing—original draft preparation, Delsi Robinsha S; writing—review and editing B.Amutha; visualization, Delsi Robinsha S; supervision, B.Amutha; All authors have read and agreed to the published version of the manuscript.

## REFERENCES


Alsharif, M. H., Kelechi, A. H., Jahid, A., Kannadasan, R., Singla, M. K., Gupta, J., & Geem, Z. W. (2024). A comprehensive survey of energy-efficient computing to enable sustainable massive IoT networks. *Alexandria Engineering Journal, 91*, 12–29. DOI: 10.1016/j.aej.2024.01.067

Bhatia, M. (2022). Energy efficient IoT-based informative analysis for edge computing environment. *Transactions on Emerging Telecommunications Technologies, 33*(9), e4527. Advance online publication. DOI: 10.1002/ett.4527

Chakraborty, C., Othman, S. B., Almalki, F. A., & Sakli, H. (2023). FC-SEEDA: Fog computing-based secure and energy efficient data aggregation scheme for Internet of healthcare Things. *Neural Computing & Applications*, *36*(1), 241–257. DOI: 10.1007/s00521-023-08270-0

Idrees, A. K., & Jawad, L. W. (2023). Energy-efficient Data Processing Protocol in edge-based IoT networks. *Annales des Télécommunications*, *78*(5–6), 347–362. DOI: 10.1007/s12243-023-00957-8

Ji, T., Luo, C., Yu, L., Wang, Q., Chen, S., Thapa, A., & Li, P. (2022). Energy-Efficient computation offloading in mobile edge computing systems with uncertainties. *IEEE Transactions on Wireless Communications*, *21*(8), 5717–5729. DOI: 10.1109/TWC.2022.3142685

Khan, S., Shah, I. A., Aurangzeb, K., Ahmad, S., Khan, J. A., & Anwar, M. S. (2024). Energy efficient task scheduling using fault tolerance technique for IoT applications in FOG computing environment. *IEEE Internet of Things Journal, 11*(24), 39009–39019. DOI: 10.1109/JIOT.2024.3403003

Khanh, Q. V., Nguyen, V., Minh, Q. N., Van, A. D., Anh, N. L., & Chehri, A. (2023). An efficient edge computing management mechanism for sustainable smart cities. *Sustainable Computing: Informatics and Systems*, *38*, 100867. DOI: 10.1016/j.suscom.2023.100867

Konecny, J., Prauzek, M., & Borova, M. (2023). Fuzzy controlled Wavelet-Based edge Computing method for Energy-Harvesting IoT sensors. *IEEE Internet of Things Journal, 10*(21), 18909–18918. DOI: 10.1109/JIOT.2023.3292915

Kong, X., Duan, G., Hou, M., Shen, G., Wang, H., Yan, X., & Collotta, M. (2022). Deep reinforcement Learning-Based Energy-Efficient edge computing for internet of vehicles. *IEEE Transactions on Industrial Informatics*, *18*(9), 6308–6316. DOI: 10.1109/TII.2022.3155162

Li, D., Lan, M., & Hu, Y. (2022). Energy-saving service management technology of internet of things using edge computing and deep learning. *Complex & Intelligent Systems*, *8*(5), 3867–3879. DOI: 10.1007/s40747-022-00666-0

Liu, D., Liang, H., Zeng, X., Zhang, Q., Zhang, Z., & Li, M. (2022). Edge computing application, architecture, and challenges in ubiquitous power internet of things. *Frontiers in Energy Research*, *10*, 850252. Advance online publication. DOI: 10.3389/fenrg.2022.850252

Liu, J., Li, C., & Luo, Y. (2024). Efficient resource allocation for IoT applications in mobile edge computing via dynamic request scheduling optimization. *Expert Systems with Applications*, *255*, 124716. DOI: 10.1016/j.eswa.2024.124716

Mahenge, M. P. J., Li, C., & Sanga, C. A. (2022). Energy-efficient task offloading strategy in mobile edge computing for resource-intensive mobile applications. *Digital Communications and Networks*, *8*(6), 1048–1058. DOI: 10.1016/j.dcan.2022.04.001

Mao, Y., You, C., Zhang, J., Huang, K., & Letaief, K. B. (2017b). A survey on Mobile Edge Computing: The Communication Perspective. *IEEE Communications Surveys and Tutorials*, *19*(4), 2322–2358. DOI: 10.1109/COMST.2017.2745201

Milani, S., Garlisi, D., Carugno, C., Tedesco, C., & Chatzigiannakis, I. (2024). Edge2LoRa: Enabling edge computing on long-range wide-area Internet of Things. *Internet of Things : Engineering Cyber Physical Human Systems*, *27*, 101266. DOI: 10.1016/j.iot.2024.101266

Minh, Q. N., Nguyen, V., Quy, V. K., Ngoc, L. A., Chehri, A., & Jeon, G. (2022). Edge computing for IoT-Enabled smart Grid: The future of energy. *Energies*, *15*(17), 6140. DOI: 10.3390/en15176140

Premalatha, B., & Prakasam, P. (2023). Optimal Energy-efficient Resource Allocation and Fault Tolerance scheme for task offloading in IoT-FoG Computing Networks. *Computer Networks*, *238*, 110080. DOI: 10.1016/j.comnet.2023.110080

Quy, N. M., Ngoc, L. A., Ban, N. T., Van Hau, N., & Quy, V. K. (2023). Edge Computing for Real-Time Internet of Things Applications: Future Internet Revolution. *Wireless Personal Communications*, *132*(2), 1423–1452. DOI: 10.1007/s11277-023-10669-w

Robinsha, S. D., & Amutha, B. (2023). IoT architecture for energy management in smart cities. *International Journal of Services Operations and Informatics*, *12*(4), 325–343. DOI: 10.1504/IJSOI.2023.137474

Robinsha, S. D., & Amutha, B. (2024). Velocious: A Resilient IoT Architecture for 6G Based Intelligent Transportation System with Expeditious Movement Mechanism. *Wireless Personal Communications*. Advance online publication. DOI: 10.1007/s11277-024-11072-9

Saadallah, N. R., & Alabady, S. A. (2024). A comprehensive study on energy-efficient-based routing protocols in the Internet of things Part I: Definition and classification. *Iran Journal of Computer Science*, *7*(3), 631–661. DOI: 10.1007/s42044-024-00180-1

Saleem, M., Shakir, M., Usman, M., Bajwa, M., Shabbir, N., Ghahfarokhi, P. S., & Daniel, K. (2023). Integrating Smart Energy Management System with Internet of Things and Cloud Computing for Efficient Demand Side Management in Smart Grids. *Energies*, *16*(12), 4835. DOI: 10.3390/en16124835

Sisi, Z., & Souri, A. (2021). Blockchain technology for energy-aware mobile crowd sensing approaches in Internet of Things. *Transactions on Emerging Telecommunications Technologies*, *35*(4), e4217. Advance online publication. DOI: 10.1002/ett.4217

Vinothkumar, T., Sivaraju, S., Thangavelu, A., & Srithar, S. (2023). An energy-efficient and reliable data gathering infrastructure using the internet of things and smart grids. *Automatika (Zagreb)*, *64*(4), 720–732. DOI: 10.1080/00051144.2023.2205724

Wadhwa, S., Rani, S., Kavita, N., Verma, S., Shafi, J., & Wozniak, M. (2022). Energy Efficient Consensus Approach of Blockchain for IoT Networks with Edge Computing. *Sensors (Basel)*, *22*(10), 3733. DOI: 10.3390/s22103733 PMID: 35632142

Xiang, Z., Zheng, Y., He, M., Shi, L., Wang, D., Deng, S., & Zheng, Z. (2021). Energy-effective artificial internet-of-things application deployment in edge-cloud systems. *Peer-to-Peer Networking and Applications*, *15*(2), 1029–1044. DOI: 10.1007/s12083-021-01273-5

Zhu, S., Ota, K., & Dong, M. (2022). Energy-Efficient artificial intelligence of things with intelligent edge. *IEEE Internet of Things Journal*, *9*(10), 7525–7532. DOI: 10.1109/JIOT.2022.3143722

# Chapter 10
# Prediction of Payment Method in Carsharing

**Antonio Perez de Juan**
https://orcid.org/0009-0003-5126-5359
*Universidad Rey Juan Carlos, Spain*

**Iñigo Martin Melero**
https://orcid.org/0000-0001-7705-6051
*University of Castilla-La Mancha, Spain*

**Raul Gomez Martinez**
https://orcid.org/0000-0003-3575-7970
*Universidad Rey Juan Carlos, Spain*

**Maria Luisa Medrano García**
https://orcid.org/0000-0003-1844-1034
*Universidad Rey Juan Carlos, Spain*

## ABSTRACT

*The rise of urban mobility platforms using VTC licenses (Uber, Cabify, Bolt) allows users to pay in cash or via registered cards/accounts. This study applies machine learning to predict the payment method (card or cash) based on income levels. After cleaning incomplete records, we compiled 280,000 VTC services, focusing on "origin postal code" and "payment method," alongside "disposable average income" from the Tax Agency's IRPF statistics. Results show 85% accuracy, with a training accuracy of 0.865 and test accuracy of 0.808. Precision reached 0.841 (train) and 0.792 (test), recall 0.885 (train) and 0.843 (test), and MCC 0.714 (train) and 0.617 (test). Findings confirm that lower-income postal codes correlate with higher cash*

DOI: 10.4018/979-8-3373-2802-7.ch010

*payments due to financial exclusion, including lack of banking access for irregular immigrants, difficulties in obtaining credit cards for low-income individuals, and account garnishments promoting informal economy reliance.*

## INTRODUCTION

The widespread adoption of smartphones among the majority of the population has led to a disruption in multiple aspects of daily life in Spain. This transformation has reshaped communication, from the early emergence of BlackBerry Messenger to the later dominance of WhatsApp (Soler, 2024), mobility with the rise of apps such as Uber, Cabify, and Bolt (Hasselwander et al., 2025), and shopping habits with the emergence of Amazon and online supermarket websites (Valadares, 2023). This technological revolution in Spain has transcended social classes, with a smartphone penetration rate of 95% (CNMC, 2023). Not having a smartphone can now be a clear source of social and financial exclusion (Cernadas Ramos et al., 2022).

The key question is whether this trend toward technological democratization has been accompanied by true financial democratization, transforming payment habits (Ferrando et al., 2023). At first glance, the shift from cash to credit or debit card payments seems undeniable due to the numerous advantages of electronic payments, such as physical security, convenience, speed, and the avoidance of carrying coins or banknotes that can be lost or deteriorate. At first glance, the answer seems clear, given the few advantages of cash payments (Gómez Fernández et al., 2020).

To understand the implications of these new mobility trends, it is important to highlight the role of public administrations in promoting alternative fuels to diesel and gasoline. Authorities have encouraged the use of hybrid, electric, hydrogen, and gas-powered vehicles by providing them access to Low Emission Zones and allowing them to operate even if they do not meet minimum vehicle or power requirements (BOE, 2015). It is also important to highlight the fundamental role of public administrations in regulating cash payments. For example, in Spain, paying more than €1,000 in cash is prohibited, in order to prevent money laundering and terrorist financing. (BOE, 2021) However, these types of regulations deprive those who have difficulty accessing electronic money from consuming goods of a certain value, extending financial exclusion and even causing technological exclusion, even in urban environments. (Alcaraz et al, 2021)

In this paper, we will analyze whether machine learning can be used to predict cash payment behavior in VTC journeys based on the average income per inhabitant of the origin postal code.

Therefore, our **hypothesis** is as follows:

1. It is possible to predict the payment method of VTC services based on the income level of the origin postal code.

## THEORETICAL FRAMEWORK

Introduction to the problem: Customer payment behavior is a key area of interest for corporations and financial services, as it impacts cash flow management and strategic decision-making within an organization (Concha et al., 2021). With the increasing availability of data, the application of machine learning techniques allows for the identification of patterns in users' payment behavior based on various factors, such as the average income of their residential area. Relationship Between Income and Payment Behavior. Income vs. Payment Preference: Some studies suggest that average income can influence payment preferences. Higher-income customers tend to use electronic payment methods (debit or credit cards) due to their access to financial products such as bank accounts and cards (World Bank, 2021). Financial Exclusion and Income: Access to basic financial products, such as checking accounts and credit or debit cards, varies significantly by income level (Somogyvári, 2021), directly impacting how individuals choose to manage their payments (Moreno de Barreda Halcón, 2021).

Machine Learning Models Applied to Payment Preference Prediction: Logistic Regression: This model is frequently used for binary classification problems (Ferrante, 2021) and could serve as an approach to predict whether a customer will pay in cash or by card. Its simplicity makes it a valuable tool for understanding how the average income of a postal code affects the likelihood of selecting a specific payment method (Pedregosa et al., 2011). Decision Trees and Random Forests: Decision trees help identify how different factors, such as income, influence the selection of a payment method. Random Forests, a more sophisticated approach, combine multiple trees to improve accuracy and effectively manage interactions between variables (Breiman, 2001). Advanced Models and Neural Networks: More complex models, such as deep neural networks, can capture nonlinear relationships between income and payment preferences. However, these models require large amounts of data and can be challenging to interpret (Goodfellow et al., 2016).

Key Variables of Interest: Average Income by Postal Code: A key variable that serves as an indicator of the socioeconomic level of an area and, potentially, the likelihood of using electronic payment methods. Payment Method: Card or cash.

Model Evaluation and Performance Metrics: In this study, we will use **accuracy**, a fundamental metric in evaluating classification models in machine learning (Zhang et al., 2020). Accuracy is defined as the proportion of correctly classified instances relative to the total instances evaluated. The formula is:

$$Accuracy = \frac{TP + TN}{TP + TN + FP + FN}$$

- **True Positives (TP):** Instances correctly classified as positive.
- **True Negatives (TN):** Instances correctly classified as negative.
- **False Positives (FP):** Instances incorrectly classified as positive.
- **False Negatives (FN):** Instances incorrectly classified as negative.

We have the following limitations: Accuracy as a Sole Metric: Relying only on accuracy can be misleading, especially in datasets with significant class imbalances. For example, if 90% of records belong to the negative class, a model that classifies everything as negative could still achieve 90% accuracy, despite being ineffective. This bias could arise in this study, considering that individuals from lower-income areas might use VTC services less frequently. However, we believe this does not significantly impact our findings.

Sensitivity to Class Imbalance: When the dataset has an uneven distribution of classes (e.g., more card payments than cash payments), a model could achieve high accuracy simply by predicting the majority class. In such cases, accuracy does not adequately reflect the model's ability to correctly identify both payment methods. To mitigate these issues, we will also consider other evaluation metrics, such as precision, recall, and the Matthews Correlation Coefficient (MCC), to provide a more balanced assessment of the model's performance.

Lack of Interpretation in High Variance Cases: Some models may exhibit inconsistent performance across different samples or regions of the feature space. A high overall accuracy does not guarantee that the model will perform equally well across all subsets or conditions of the data, such as postal codes with medium income compared to those with high or low income.

Potential Disconnect with Result Interpretation: In socioeconomic studies, it is important not only to make accurate predictions but also to understand why a certain payment preference occurs. Accuracy does not provide insights into the contribution of each variable, which limits the ability to interpret results and derive practical implications.

After analyzing the above limitations, we conclude that they are not significant, as the sample size is large enough to mitigate them, and the service is transversal across virtually all social strata.

We will accept the hypothesis if accuracy > 0.5, as this indicates a correlation between the variables.

Practical Applications and Benefits: Service Optimization: Accurate prediction of the payment method can help businesses improve their cash management and reduce operational costs. Personalization of Services: Data on payment behavior can be used to offer personalized promotions or loyalty programs based on customer preferences (Garcia, 2024).

## METHODOLOGY

A company operating in the VTC sector has provided a database of services performed through the "Bolt" platform between 20/05/2022 and 21/03/2024, with the following fields: Driver ID, Driver Name, Invoice ID, Order ID, State, Address, Destination Address, Order Created Time, Driver Arrived to Pickup, Order Ride Start Time, Order Ride End Time, Balance Entry Date, Payment Method Type, Type, Trip Payments, Compensated Cash, Discounts, Gross Earnings, Tips, Toll Road, Fees Net, Earnings, Personal Code, Compensated In App, Discounts.

Additionally, data from the Income Tax Declaration Statistics of the largest municipalities published by the Tax Agency is available, which will allow us to cross-reference both datasets, where we have the variables Postal Code and Average Disposable Income.

From the fields mentioned above, for this study, we will focus only on "Payment Method Type" and "Address" from the first table. From the second table, we will use both fields (Postal Code and Average Disposable Income).

After filtering the database of approximately 500,000 records, we select those containing a postal code that appears in the Income Tax Declaration Statistics of the largest municipalities by postal code (3). This reduces the sample to 282,770 valid records. Furthermore, we extract the postal code from the "Address" field.

To predict payment behavior, we will model the "Payment Method Type" variable as 1 or 0, which is later encoded as "Payment Binary" in the subsequent table.

*Table 1. Dataset*

| Row | Trip Payments | Postal Code income | ... | Payment Binary |
|---|---|---|---|---|
| 0 | 13.40 | 35673 | ... | 0 |
| 1 | 19.55 | 35279 | ... | 1 |
| 2 | 3.60 | 43380 | ... | 1 |
| 3 | 5.05 | 41479 | ... | 1 |
| ... | ... | ... | ... | ... |
| 282769 | 9.00 | 19834 | ... | 0 |
| 282770 | 4.75 | 35279 | ... | 1 |

Afterward, using machine learning, the goal is to achieve an accuracy greater than 0.5, which indicates that prediction is possible. In this case, as we can see in the following table, the accuracy rate is 85%.

*Table 2. Results of the study*

| | Train set | Test set |
|---|---|---|
| Accuracy | 0.856 | 0.808 |
| Precision | 0.841 | 0.792 |
| Recall | 0.885 | 0.843 |
| MCC | 0.714 | 0.617 |

## CONCLUSIONS

This study demonstrates the capability of machine learning to identify patterns in payment decisions in urban mobility services based on socio-economic variables. Using a dataset of VTC trips and employing classification models such as logistic regression and Random Forests, an accuracy of 85% was achieved in predicting the payment method (cash or card) based on the average income of the postal code of origin. This claim is substantiated by the following results: accuracy of 0.865 in the training set and 0.808 in the test set, precision of 0.841 in the training set and 0.792 in the test set, recall of 0.885 in the training set and 0.843 in the test set, and MCC of 0.714 in the training set and 0.617 in the test set. The analysis confirms that average income is a significant predictor, associated with a higher likelihood of cash payments in more disadvantaged areas.

The implementation of machine learning algorithms has allowed for the validation of the hypothesis that financial exclusion and the informal economy influence payment methods. Factors such as migratory status, limited access to banking services, and credit restrictions encourage reliance on cash in these areas. Specifically, areas with higher cash payment rates are the working-class neighborhoods of Madrid, which have higher percentages of irregular migrants and low-income populations.

The results of this study can guide both VTC platforms and those responsible for preventing financial exclusions. On one hand, the accurate prediction of payment methods based on income allows businesses to optimize cash management, reduce operational costs, and tailor their services to user preferences. On the other hand, identifying socio-economic factors linked to cash usage highlights the need to strengthen financial inclusion initiatives for low-income populations, facilitating access to banking services and reducing cash dependency.

In conclusion, this work emphasizes the usefulness of artificial intelligence in predicting financial behaviors and understanding their socio-economic determinants, providing a reference framework for both the private sector and the design of financial inclusion policies.

## FUTURE LINES OF RESEARCH

### Expansion of Socio-Economic Variables

Future research could explore the inclusion of additional socio-economic variables, such as education level or digital literacy in different areas, to assess their impact on payment method preferences in VTC services. Understanding how factors like educational background or technological adoption correlate with payment choices would provide valuable insights into socio-economic determinants of financial behavior. Temporal Segmentation: Investigating seasonal or daily variations in cash vs. card payment usage could further refine predictive models. For instance, payment patterns might differ at certain times of the day, days of the week, or even during specific months (e.g., holidays or economic events). This time-based segmentation could enhance the accuracy of predictions and allow for more dynamic strategies tailored to different times.

### Segmentation by Transaction Amount

Exploring whether payment method choices correlate with the amount of the transaction could be another area for study. It would be useful to determine if larger or smaller ticket sizes influence the preference for cash or card payments, as people may choose cash for smaller amounts and cards for larger ones.

### Application of Explainable AI Models

One limitation of current machine learning models is their lack of interpretability. Developing explainable AI models that provide clear insights into how specific socio-economic factors influence payment decisions could help policymakers and businesses better understand the key drivers of behavior and make more informed decisions regarding financial inclusion strategies. *International Comparison*: Conducting comparative studies across different countries or regions could reveal how payment behavior varies in relation to income, cultural differences, and economic environments. For example, studying countries with high financial inclusion versus those with lower rates could provide a broader perspective on how socio-economic factors impact payment choices globally. These additional research lines could lead to a more comprehensive understanding of the dynamics at play in payment behaviors, offering valuable insights for businesses, policymakers, and financial inclusion advocates.

## REFERENCES


Alcaraz, M. C., Monteagudo, I. C., & Tormo, F. V. S. (2021). The contribution of credit cooperatives to alleviating geographic financial exclusion in less populated areas: The case of Spain. *CIRIEC-Spain, Journal of Public, Social, and Cooperative Economy*, (103), 197–224.

BOE (2015). Royal Decree 1057/2015, of November 20, which modifies the Regulation of the Law on the Organization of Land Transport, approved by Royal Decree 1211/1990, of September 28, regarding the leasing of vehicles with driver, to adapt it to Law 9/2013, of July 4, which modifies Law 16/1987, of July 30, on the Organization of Land Transport and Law 21/2003, of July 7, on Air Safety.

BOE (2021). Law 11/2021, of July 9, on measures for the prevention and fight against tax fraud, transposing Directive (EU) 2016/1164 of the Council, of July 12, 2016, establishing rules against tax avoidance practices that directly affect the functioning of the internal market, amending various tax regulations and matters relating to gambling regulation.

Breiman, L. (2001). Random Forests. *Machine Learning*, *45*(1), 5–32. DOI: 10.1023/A:1010933404324

Cernadas Ramos, A., Barral Buceta, B., & Fernández Da Silva, Á. (2022). Digital divide and social exclusion: Can ICT change the status quo? *Revista Brasileira de Políticas Públicas, 12*(2).

Concha, E. J. T., & Reyes, V. M. L. (2021). Over-indebtedness and its impact on credit risk in Peruvian microfinance institutions: Municipal savings and credit banks 2016–2019. *Contabilidad y Negocios*, *16*(32), 78–90.

Ferrando, L., & Posada, D. (2023). Use of cash and other payment methods: How is the way we pay changing? *Economic Bulletin - Bank of Spain, 2023*(Q1), 01. https://doi.org/DOI: 10.53479/2499

Ferrante. (2021). Artificial intelligence and algorithmic biases. *Nueva Sociedad*.

Goodfellow, I., Bengio, Y., & Courville, A. (2016). *Deep Learning*. MIT Press.

Hasselwander, M., Lütjens, K., Scheier, B., Frieske, B., Ennen, D., Scheelhaase, J., Kumar, S., & Maertens, S. (2025). Digital platforms, multimodal transport, and super apps. *Transportation Research Procedia*, *86*, 207–212. DOI: 10.1016/j.trpro.2025.04.027

Pedregosa, F., Varoquaux, G., & Gramfort, A. (2011). Scikit-learn: Machine learning in Python. *Journal of Machine Learning Research, 12*, 2825–2830.

Somogyvári, M. (2021) Financial Exclusion in the Digital Payment Space, *Financial and Economic Review*, (65-86)

Valadares, G. C., de Rezende, D. C., Carvalho, J. F. S., dos Reis, N. D., Evangelista, J. M. F., & Costa, A. P. (2023). Smartphone use and purchasing behavior in supermarkets: An approach based on Practice Theory. *Revista da UI_IPSantarém, 11*(4), 64–76.

World Bank. (2021). *Global Findex Database 2021: Financial Inclusion, Digital Payments, and Resilience in the Age of COVID-19*. Retrieved from https://www.worldbank.org

# Compilation of References


Adams, C., Pente, P., Lemermeyer, G., Turville, J., & Rockwell, G. (2022). Artificial Intelligence and Teachers' New Ethical Obligations. *International Journal of Information Ethics, 31*(1). Advance online publication. DOI: 10.29173/irie483

Adeyele, V. O., & Ramnarain, U. (2024). Exploring the Integration of ChatGPT in Inquiry-Based Learning: Teacher Perspectives. *International Journal of Technology in Education*, *7*(2), 200–217. DOI: 10.46328/ijte.638

Agüero, A., Bustelo, M., y Viollaz, M. (2020). Inequalities in the digital world?: Gender gaps in the use of ICT. https://doi.org/DOI: 10.18235/0002235

Aguirre, E., Zhang, H., & Liu, F. (2022). Privacy concerns in the era of autonomous vehicles: Data security and ethical challenges for advertisers. *Journal of Media Ethics*, *16*(3), 151–165.

Ahn, D., & Lim, H. (2025). Exploring K-12 Physical Education Teachers' Perspectives on Opportunities and Challenges of AI Integration Through Ideation Workshops. *CHI Conference on Human Factors in Computing Systems (CHI '25),* April 26-May 1, 2025*,* Yokohama, Japan*, 1*. DOI: 10.1145/3706598.3713646

Alabi, O., & Bakare, A. (2019). Application of IoT in Smart Waste Management Systems. *International Journal of Environmental Science and Technology*, *16*(5), 2541–2552.

Alam, A., & Mohanty, A. (2022). Foundation for the Future of Higher Education or 'Misplaced Optimism'? Being Human in the Age of Artificial Intelligence. Paper presented at the *International Conference on Innovations in Intelligent Computing and Communications,* 17–29. https://doi.org/DOI: 10.1007/978-3-031-23233-6_2

Albino, V., Berardi, U., & Dangelico, R. M. (2015). Smart Cities: Definitions, Dimensions, Performance, and Initiatives. *Journal of Urban Technology*, *22*(1), 3–21. DOI: 10.1080/10630732.2014.942092

Alcaraz, M. C., Monteagudo, I. C., & Tormo, F. V. S. (2021). The contribution of credit cooperatives to alleviating geographic financial exclusion in less populated areas: The case of Spain. *CIRIEC-Spain, Journal of Public, Social, and Cooperative Economy*, (103), 197–224.

Alfaro, L., Kalemli-Ozcan, S., & Volosovych, V. (2008). Why doesn't capital flow from rich to poor countries? An empirical investigation. *The Review of Economics and Statistics*, *90*(2), 347–368. DOI: 10.1162/rest.90.2.347

Alkhalil, A., Abdallah, M. A. E., Alogali, A., & Aljaloud, A. (2021). Applying Big Data Analytics in Higher Education: A Systematic Mapping Study. [IJICTE]. *International Journal of Information and Communication Technology Education*, *17*(3), 29–51. DOI: 10.4018/IJICTE.20210701.oa3

Almenara, J. C. (2014). Reflexiones sobre la brecha digital y la educación: Siguiendo el debate. *Inmanencia*, *4*(2), 14–46.

Alpuche Ortiz, C. M., González Torres, J., & Corona Ferreira, A. (2024). Plataforma Educativa para el Acondicionamiento en Línea de Educación Física. *Ciencia Latina Revista Científica Multidisciplinar*, *8*(4), 13643–13656. DOI: 10.37811/cl_rcm.v8i4.13868

Alsharif, M. H., Kelechi, A. H., Jahid, A., Kannadasan, R., Singla, M. K., Gupta, J., & Geem, Z. W. (2024). A comprehensive survey of energy-efficient computing to enable sustainable massive IoT networks. *Alexandria Engineering Journal*, *91*, 12–29. DOI: 10.1016/j.aej.2024.01.067

Alyami, S. H., & Rezgui, Y. (2012). Sustainable building assessment tool development approach. *Sustainable Cities and Society*, *5*, 52–62. DOI: 10.1016/j.scs.2012.05.004

Al-Zahrani, A. M. (2024). Unveiling the shadows: Beyond the hype of AI in education. *Heliyon*, *10*(9), e30696. DOI: 10.1016/j.heliyon.2024.e30696 PMID: 38737255

Ammad, U., Ahmad, A., & Qamar, F. (2020). Autonomous public transportation in smart cities: A case study of RoboBusLine in Helsinki. *International Journal of Transportation Science and Technology*, *9*(4), 283–294.

Anderson, J. M., Kalra, N., Stanley, K. D., Sorensen, P., Samaras, C., & Oluwatola, O. A. (2018). *Autonomous vehicle technology: A guide for policymakers*. RAND Corporation.

Ashby, W. R. (2018). *An Introduction to Cybernetics*. Creative Media Partners, LLC.

Ashley, C., & Mitchell, J. (2009). *Tourism and poverty reduction: Pathways to prosperity*. Routledge. DOI: 10.4324/9781849774635

Asociación Española de Psiquiatría de la Infancia y la Adolescencia. (2024). *Recomendaciones de Uso de Nuevas Tecnologías en la Infancia y Adolescencia.*

Ayyildiz, P., & Yilmaz, A. (2023). A New Chapter Is Being Written about Writing Instruction: Instructional Leadership at K-12 Levels in the Age of Artificial Intelligence (AI). *Educational Policy Analysis and Strategic Research, 18*(4), 82–101. DOI: 10.29329/epasr.2023.631.4

Aznar-Minguet, P., Ull, M. Á., Martínez-Agut, M. P., & Piñero, A. (2017). Evaluar para transformar: evaluación de la docencia universitaria bajo el prisma de la sostenibilidad. Enseñanza de las ciencias: revista de investigación y experiencias didácticas, 35(1), 5-27.

Bakıcı, T., Almirall, E., & Wareham, J. (2013). A smart city initiative: The case of Barcelona. *Journal of the Knowledge Economy, 4*(2), 135–148. DOI: 10.1007/s13132-012-0084-9

Banerjee, S. B. (2020). *Corporate social responsibility: The good, the bad, and the ugly*. Edward Elgar Publishing.

Barro, S. (2025). The Digital Society Is More Unequal. Información Comercial Española, ICE. *Journal of Economics*, (938), 97–111. DOI: 10.32796/ice.2025.938.7890

Barua, S. (2020). Financing sustainable development goals: A review of challenges and mitigation strategies. *Business Strategy & Development, 3*(3), 277–293. DOI: 10.1002/bsd2.94

Batty, M., Axhausen, K. W., Giannotti, F., Pozdnoukhov, A., Bazzani, A., Wachowicz, M., Ouzounis, G., & Portugali, Y. (2012). Smart cities of the future. *The European Physical Journal. Special Topics, 214*(1), 481–518. DOI: 10.1140/epjst/e2012-01703-3

Beer, S. (1979). *The Heart of Enterprise*. Wiley.

Beer, S. (1981). *Brain of the firm: the managerial cybernetics of organization*. Wiley.

Beer, S. (1985). *Diagnosing the System for Organizations*. Wiley.

Belz, F. M., & Peattie, K. (2012). *Sustainability marketing: A global perspective*. John Wiley & Sons.

Benzizoune, O. (2024). Enhancing Classroom Management and Student Engagement: The Role of ClassDojo and Gamification in Education. *Journal of English Language Teaching and Applied Linguistics, 6*(3), 106–114. DOI: 10.32996/jeltal.2024.6.3.13

Bhatia, M. (2022). Energy efficient IoT-based informative analysis for edge computing environment. *Transactions on Emerging Telecommunications Technologies, 33*(9), e4527. Advance online publication. DOI: 10.1002/ett.4527

Bitzer, D. L., Lyman, E. R., & Easley, A. J. (1965). *The Uses Of Plato: A Computer Controlled Teaching System.* DOI: 10.21236/AD0623526

Bloomberg, N. E. F. (2021). *Electric vehicle outlook 2021.* Bloomberg New Energy Finance.

Bodemer, O. (2023). *Enhancing Individual Sports Training through Artificial Intelligence: A Comprehensive Review.* DOI: 10.36227/techrxiv.24005916

BOE (2015). Royal Decree 1057/2015, of November 20, which modifies the Regulation of the Law on the Organization of Land Transport, approved by Royal Decree 1211/1990, of September 28, regarding the leasing of vehicles with driver, to adapt it to Law 9/2013, of July 4, which modifies Law 16/1987, of July 30, on the Organization of Land Transport and Law 21/2003, of July 7, on Air Safety.

BOE (2021). Law 11/2021, of July 9, on measures for the prevention and fight against tax fraud, transposing Directive (EU) 2016/1164 of the Council, of July 12, 2016, establishing rules against tax avoidance practices that directly affect the functioning of the internal market, amending various tax regulations and matters relating to gambling regulation.

Bofill, J., Pla-Campas, G., & Sebastiani, E. M. (2025). Is Artificial Intelligence an educational resource in Physical Education? A Systematic Review. *Apunts. Educacion Fisica y Deportes, 160*, 1–9. https://doi.org/.(2025/2).160.01DOI: 10.5672/APUNTS.2014-0983.ES

Boley, B. B., & Uysal, M. (2013). Competitive synergy through practicing triple bottom line sustainability: Evidence from three hospitality case studies. *Tourism and Hospitality Research, 13*(4), 226–238. DOI: 10.1177/1467358414528528

Bond, M., Khosravi, H., De Laat, M., Bergdahl, N., Negrea, V., Oxley, E., Pham, P., Chong, S. W., & Siemens, G. (2024). A Meta Systematic Review of Artificial Intelligence in Higher Education: A Call for Increased Ethics, Collaboration, and Rigour. *International Journal of Educational Technology in Higher Education, 21*(1), 4–41. DOI: 10.1186/s41239-023-00436-z

Borenstein, J., Herkert, J. R., & Herfeld, C. (2020). *The ethics of autonomous cars.* The Atlantic. https://www.theatlantic.com

Bourn, D. (2020). *Understanding Global Skills for 21st Century Professions.* Palgrave Macmillan., DOI: 10.1007/978-3-030-44154-1

Bowen, F. (2014). *After greenwashing: Symbolic corporate environmentalism and society*. Cambridge University Press. DOI: 10.1017/CBO9781139541213

Bowen, F., & Aragon-Correa, J. A. (2014). Greenwashing in corporate environmentalism research and practice: The importance of what we say and do. *Organization & Environment, 27*(2), 107–112. DOI: 10.1177/1086026614537078

Bozkurt, A. (2023). Unleashing the Potential of Generative AI, Conversational Agents and Chatbots in Educational Praxis: A Systematic Review and Bibliometric Analysis of GenAI in Education. *Open Praxis, 15*(4), 261–270. DOI: 10.55982/openpraxis.15.4.609

Brands Called Out For Greenwashing In 2020 - CWR - China Water Risk. (2020, diciembre 17). *China Water Risk*. Retrieved from https://cwrrr.org/opinions/8-brands-called-out-for-greenwashing-in-2020/

Brantingham, P. J., Valasik, M., & Mohler, G. (2018). Does predictive policing lead to biased arrests? Results from a randomized controlled trial. *Statistics and Public Policy (Philadelphia, Pa.), 5*(1), 1–6. DOI: 10.1080/2330443X.2018.1438940

Breetz, H. L., Mccormick, D. M., & Nykvist, B. (2021). Electric vehicle adoption in the United States: Influences of policy and environmental attitudes. *Journal of Environmental Economics and Policy, 7*(3), 276–291.

Breiman, L. (2001). Random Forests. *Machine Learning, 45*(1), 5–32. DOI: 10.1023/A:1010933404324

Brown, A., & Smith, L. (2023). The consequences of corporate greenwashing on consumer trust. *Journal of Sustainable Marketing, 11*(2), 45–60. DOI: 10.1234/jsm.v11i2.456

Buchanan, C., Howitt, M. L., Wilson, R., Booth, R. G., Risling, T., & Bamford, M. (2021). Predicted Influences of Artificial Intelligence on Nursing Education: Scoping Review. *JMIR Nursing, 4*(1), e23933. DOI: 10.2196/23933

Buckley, G. P. (Ed.). (2012). *Ecology and management of coppice woodlands*. Springer Science & Business Media.

Burr, C., Taddeo, M., & Floridi, L. (2020). The Ethics of Digital Well-Being: A Thematic Review. *Science and Engineering Ethics, 26*(4), 2313–2343. DOI: 10.1007/s11948-020-00175-8

California Air Resources Board. (2021). *California's 2021 zero-emission vehicle regulations*.

Camilleri, M. A. (2022). Walking the talk about corporate social responsibility communication: An elaboration likelihood model perspective. *Business Ethics, the Environment & Responsibility, 31*(3), 649–661. DOI: 10.1111/beer.12427

Cano, L. P. (2014). La corrupción y la ineficiencia en el gasto público local y su impacto en la pobreza en Colombia. http://hdl.handle.net/11445/712

Carbajo Blanco, A. (2021). Ecoembes: Implementing innovation and social responsibility strategies.

Carbon Market Watch. (2022). *Offsetting aviation emissions: The illusion of carbon neutrality*. https://carbonmarketwatch.org/publications

Carbon Market Watch. (2023). *Hospitality and the climate crisis: Why offsetting isn't enough*. https://carbonmarketwatch.org/publications

Cardona, T., Cudney, E. A., Hoerl, R., & Snyder, J. (2023). Data Mining and Machine Learning Retention Models in Higher Education. *Journal of College Student Retention, 25*(1), 51–75. DOI: 10.1177/1521025120964920

Carlson, L., Grove, S. J., & Kangun, N. (1993). A content analysis of environmental advertising claims: A matrix method approach. *Journal of Advertising, 22*(3), 27–39. DOI: 10.1080/00913367.1993.10673409

Cernadas Ramos, A., Barral Buceta, B., & Fernández Da Silva, Á. (2022). Digital divide and social exclusion: Can ICT change the status quo? *Revista Brasileira de Políticas Públicas, 12*(2).

Chakraborty, C., Othman, S. B., Almalki, F. A., & Sakli, H. (2023). FC-SEEDA: Fog computing-based secure and energy efficient data aggregation scheme for Internet of healthcare Things. *Neural Computing & Applications, 36*(1), 241–257. DOI: 10.1007/s00521-023-08270-0

Chan, C. K. Y., & Hu, W. (2023). Students' Voices on Generative AI: Perceptions, Benefits, and Challenges in Higher Education. *International Journal of Educational Technology in Higher Education, 20*(1), 43–18. DOI: 10.1186/s41239-023-00411-8

Cheah, Y. H., Lu, J., & Kim, J. (2025). Integrating Generative Artificial Intelligence in K-12 Education: Examining Teachers' Preparedness, Practices, and Barriers. *Computers and Education: Artificial Intelligence, 810*, 100363. Advance online publication. DOI: 10.1016/j.caeai.2025.100363

Chen, Y., Li, C., Wang, W., Zang, Y., Chen, X. M., & Gao, Z. (2025). The landscape, trends, challenges, and opportunities of sustainable mobility and transport. *npj Sustainable Mobility and Transport, 2*(8). DOI: 10.1038/s44333-025-00026-8

Chen, Z. (2023). Ethics and discrimination in artificial intelligence-enabled recruitment practices. *Humanities and Social Sciences Communications 2023 10:1, 10*(1), 1–12. DOI: 10.1057/s41599-023-02079-x

Chen, Y. S., & Chang, C. H. (2013). Greenwash and green trust: The mediation effects of green consumer confusion and green perceived risk. *Journal of Business Ethics, 114*(3), 489–500. DOI: 10.1007/s10551-012-1360-0

Chiu, T. K. F., Xia, Q., Zhou, X., Chai, C. S., & Cheng, M. (2023). Systematic Literature Rreview on Opportunities, Challenges, and Future Research Recommendations of Artificial Intelligence in Education. *Computers and Education. Artificial Intelligence, 4*, 100118. DOI: 10.1016/j.caeai.2022.100118

CIMAR. (2022). La brecha digital en el alumnado inmigrante: cifras según el informe PISA 2018. Congreso Internacional de Marketing, Comunicación y Estrategias. (CIMAR). https://2022.cimar.org.es/ponencia/la-brecha-digital-en-el-alumnado-inmigrante-cifras-segun-el-informe-pisa-2018/

Cirera, X., & Maloney, W. F. (2017). *La paradoja de la innovación: Las capacidades de los países en desarrollo y la promesa incumplida de la convergencia tecnológica.* Publicaciones del Banco Mundial.

City of Helsinki. (2020). *Helsinki 3D+ Digital Twin Project.* https://www.hel.fi

Clarke, R. (2019). Surveillance in the UK public sector. *Computer Law & Security Review, 35*(4), 105332.

Collie, R. J., & Martin, A. J. (2024). Teachers' Motivation and Engagement to Harness Generative AI for Teaching and Learning: The Role of Contextual, Occupational, and Background Factors. *Computers and Education. Artificial Intelligence, 6*, 100224. DOI: 10.1016/j.caeai.2024.100224

Concha, E. J. T., & Reyes, V. M. L. (2021). Over-indebtedness and its impact on credit risk in Peruvian microfinance institutions: Municipal savings and credit banks 2016–2019. *Contabilidad y Negocios, 16*(32), 78–90.

Córdova-Esparza, D., Romero-González, J., Córdova-Esparza, K., Terven, J., & López-Martínez, R. (2024). Active Learning Strategies in Computer Science Education: A Systematic Review. *Multimodal Technologies and Interaction, 8*(6), 50. DOI: 10.3390/mti8060050

Crane, A. (2000). Facing the backlash: Green marketing and strategic ambiguity. *Journal of Strategic Marketing, 8*(1), 1–17.

Crawford, J., Vallis, C., Yang, J., Fitzgerald, R., & O'Dea, C. (2023). Artificial Intelligence is awesome, but good teaching should always come first. *Journal of University Teaching & Learning Practice*, *20*(7), 1–12. DOI: 10.53761/1.20.01.01

Creemers, R. (2018). China's social credit system: An evolving practice of control. SSRN *Electronic Journal*. DOI: 10.2139/ssrn.3175792

Crompton, H., & Burke, D. (2023). Artificial Intelligence in Higher Education: The State of the Field. *International Journal of Educational Technology in Higher Education*, *20*(1), 22. DOI: 10.1186/s41239-023-00392-8

Crossrail Ltd. (2020). *Crossrail Annual Report*. https://www.crossrail.co.uk

Cukurova, M., & Miao, F. (2024). *AI Competency Framework for Teachers*. UNESCO Publishing.

Dahl, D. W. (2010). Understanding the role of marketing in consumer creativity. *Journal of Marketing*, *74*(5), 15–30. DOI: 10.1509/jmkg.74.5.15

Dahlén, M., Lange, F., & Smith, T. (2010). *Marketing communications: A brand narrative approach*. John Wiley & Sons.

Datta, A. (2018). The digital turn in postcolonial urbanism: Smart citizenship in the making of India's 100 smart cities. *Transactions of the Institute of British Geographers*, *43*(3), 405–419. DOI: 10.1111/tran.12225

David, S., Bačić, B., Richter, C., & Mundt, M. (2023). Editorial: Artificial intelligence to enhance biomechanical modelling. *Frontiers in Sports and Active Living*, *5*, 1188035. DOI: 10.3389/fspor.2023.1188035 PMID: 37188071

Davies, I. E. E., Nwankwo, C. O., Olofinnade, O. M., & Michaels, T. A. (2019, November). Insight review on impact of infrastructural development in driving the SDGs in developing nations: A case study of Nigeria. In IOP Conference Series: Materials Science and Engineering (Vol. 640, No. 1, p. 012112). IOP Publishing. https://doi.org/DOI: 10.1088/1757-899X/640/1/012112

De Oteyza, M. O. G., Sosa, M. J., & Lara, M. C. F. (2021). Retos para la integración de las personas inmigrantes en las Smart Cities: El caso de Madrid. *Cuadernos Geográficos*, *60*(1), 244–262. DOI: 10.30827/cuadgeo.v60i1.13733

Debnath, A. K., Chin, H. C., Haque, M. M., & Yuen, B. (2014). A METHODOLOGICAL FRAMEWORK FOR BENCHMARKING SMART TRANSPORT CITIES. *Cities (London, England)*, *37*, 47–56. https://vuir.vu.edu.au/34171/. DOI: 10.1016/j.cities.2013.11.004

Dehbozorgi, R., Zangeneh, S., Khooshab, E., Nia, D. H., Hanif, H. R., Samian, P., Yousefi, M., Hashemi, F. H., Vakili, M., Jamalimoghadam, N., & Lohrasebi, F. (2025). The application of artificial intelligence in the field of mental health: A systematic review. *BMC Psychiatry*, *25*(1), 1–20. DOI: 10.1186/s12888-025-06483-2 PMID: 39953464

del Álamo, J. B., Albero, C. M., Hidalgo, D. A., & Bastida, J. M. G. (2017). *Education for sustainability in Spain: reflections and proposals*. Madrid, Spain: Fundación alternativas.

Delmas, M. A., & Burbano, V. C. (2011). The drivers of greenwashing. *California Management Review*, *54*(1), 64–87. DOI: 10.1525/cmr.2011.54.1.64

Delmas, M. A., Lyon, T. P., & Maxwell, J. W. (2019). Understanding the role of information disclosure and environmental performance. *Organization & Environment*, *32*(3), 307–328. DOI: 10.1177/1086026619850185

Díaz, A. A. (2024). Implicación docente en educación Primaria sobre la Educación Ambiental. *Journal of Environmental Education and Sustainability*, *6*(1), 1203–1203.

Díaz, B., & Nussbaum, M. (2024). Artificial Intelligence for Teaching and Learning in Schools: The Need for Pedagogical Intelligence. *Computers & Education*, *217*, 105071. DOI: 10.1016/j.compedu.2024.105071

Díaz-Nafría, J. M., & Guarda, T. (2018). Participatory Sensing in Sustainable Mobility: biciLAB Model. In Á. Rocha & T. Guarda (Eds.), *Developments and Advances in Defense and Security: Proceedings of the Multidisciplinary International Conference of Research Applied to Defense and Security (MICRADS 2018)* (Vol. 94, pp. 97-104). Springer International Publishing. DOI: 10.1007/978-3-319-78605-6_8

Dignum, V. (2021). The Role and Challenges of Education for Responsible AI. *London Review of Education*, *19*(1), 1–11. DOI: 10.14324/LRE.19.1.01

Dodds, R., & Holmes, M. R. (2011). Sustainability in Canadian B&Bs: Comparing the east versus west. *International Journal of Tourism Research*, *13*(5), 482–495. DOI: 10.1002/jtr.822

Dodds, R., Holmes, M., & Novotny, M. (2022). Because I believe in it: Examining intrinsic and extrinsic motivations for sustainability in festivals through self-determination theory. *Tourism Recreation Research*, *47*(2), 111–129. DOI: 10.1080/02508281.2020.1841375

du Boulay, B. (2019). Escape from the Skinner Box: The Case for Contemporary Intelligent Learning Environments. *British Journal of Educational Technology*, *50*(6), 2902–2919. DOI: 10.1111/bjet.12860

Duan, H., & Zhao, W. (2024). The Effects of Educational Artificial Intelligence-Powered Applications on Teachers' Perceived Autonomy, Professional Development for Online Teaching, and Digital Burnout. *International Review of Research in Open and Distance Learning, 25*(3), 57–76. https://doi.org/. v25i3.7659.DOI: 10.19173/irrodl

Dwyer, L. (2018). Saluting while the ship sinks: The necessity for tourism paradigm change. *Journal of Sustainable Tourism, 26*(1), 29–48. DOI: 10.1080/09669582.2017.1308372

Earthsight. (2023). *Greenwashing in online travel platforms: A critical investigation*. https://earthsight.org

Easterly, W., & Levine, R. (2001). What have we learned from a decade of empirical research on growth? It's Not Factor Accumulation: Stylized Facts and Growth Models. *The World Bank Economic Review, 15*(2), 177–219. DOI: 10.1093/wber/15.2.177

Eastman, C., Teicholz, P., Sacks, R., & Liston, K. (2011). *BIM Handbook: A Guide to Building Information Modeling*. Wiley.

Eltahir, M. E., & Babiker, F. M. E. (2024). The Influence of Artificial Intelligence Tools on Student Performance in e-Learning Environments: Case Study. *Electronic Journal of e-Learning, 22*(9), 91–110. DOI: 10.34190/ejel.22.9.3639

Encabo, S. O. (2017). Digital Divide, Poverty, and Social Exclusion. *Labor Issues: Andalusian Journal of Work and Social Welfare*, (138), 285–313.

*Environmental Education Action Plan for Sustainability. PAEAS* (n. d.). Ministry for Ecological Transition and the Demographic Challenge. https://www.miteco.gob.es/es/ceneam/plan-accion- educacion-ambiental.html

Ertmer, P. A. (1999). Addressing First- and Second-Order Barriers to Change: Strategies for Technology Integration. *Educational Technology Research and Development, 47*(4), 47–61. DOI: 10.1007/BF02299597

Espejo, R., & Reyes, A. (2014). *Organizational Systems: Managing Complexity with the Viable System Model*. Springer Berlin Heidelberg.

Espinosa, A., & Martinez, A. (2025). The Viable System Model to Support Sustainable SelfGovernance in Communities: Learning from Case Studies. *Systemic Practice and Action Research, 38*(14), 14. Advance online publication. DOI: 10.1007/s11213-025-09724-3

European Commission. (2023). *Sustainable consumption survey: Perceptions and behaviors of European consumers*. https://europa.eu

European Commission. European Executive Agency for Education and Culture, Parveva, T., Birch, P., Horváth, A., Piedrafita Tremosa, S., et al., (2024). *Learning for sustainability in Europe: building competences and support for teachers and schools*, Birch, P.(editor), Publications Office of the European Union. https://data.europa.eu/doi/10.2797/521213

Fagerberg, J., & Srholec, M. (2008). National innovation systems, capabilities and economic development. *Research Policy*, *37*(9), 1417–1435. DOI: 10.1016/j.respol.2008.06.003

Fang, X., Misra, S., Xue, G., & Yang, D. (2012). Smart grid—The new and improved power grid: A survey. *IEEE Communications Surveys and Tutorials*, *14*(4), 944–980. DOI: 10.1109/SURV.2011.101911.00087

Feng, Y., Zhang, Y., & Liu, J. (2023). In-car advertising: The evolution of media consumption in autonomous vehicles. *Journal of Media Innovation*, *35*(2), 109–123.

Feng, Y., Zhang, Y., & Liu, J. (2023). Leveraging augmented reality for interactive advertising in autonomous vehicles. *Journal of Advertising Technology*, *38*(4), 79–93.

Ferrando, L., & Posada, D. (2023). Use of cash and other payment methods: How is the way we pay changing? *Economic Bulletin - Bank of Spain, 2023*(Q1), 01. https://doi.org/DOI: 10.53479/2499

Ferrante. (2021). Artificial intelligence and algorithmic biases. *Nueva Sociedad.*

Ferriz-Valero, A., Østerlie, O., García Martínez, S., & García-Jaén, M. (2020). Gamification in Physical Education: Evaluation of Impact on Motivation and Academic Performance within Higher Education. *International Journal of Environmental Research and Public Health, 17*(12), 4465. DOI: 10.3390/ijerph17124465 PMID: 32575919

FHWA. (2020). Adaptive Signal Control Technologies. *U.S. Federal Highway Administration*. https://www.fhwa.dot.gov

Font, X., Elgammal, I., & Lamond, I. (2017). Greenhushing: The deliberate under communicating of sustainability practices by tourism businesses. *Journal of Sustainable Tourism*, *25*(7), 1007–1023. DOI: 10.1080/09669582.2016.1158829

Font, X., & Elgazzar, S., &лом, J. (2020). Greenwashing and beyond: The need for clear, transparent and evidence-based communication about tourism sustainability. *Journal of Sustainable Tourism*, *28*(7), 959–977.

Font, X., & Lynes, J. (2018). Corporate social responsibility in tourism and hospitality. *Journal of Sustainable Tourism, 26*(7), 1027–1042. DOI: 10.1080/09669582.2018.1488856

Font, X., & McCabe, S. (2017). Sustainability and marketing in tourism: Its contexts, paradoxes, approaches, challenges and potential. *Journal of Sustainable Tourism, 25*(7), 869–883. DOI: 10.1080/09669582.2017.1301721

Font, X., & McCabe, S. (Eds.). (2018). *Marketing for sustainable tourism*. Routledge.

Freund, K., Bayne, A., Beck, L., Siegfried, A., Warren, J., Nadel, T., & Natajaran, A. (2020). Characteristics of ride share services for older adults in the United States. *Journal of Safety Research, 72*, 9-19. DOI: 10.1016/j.jsr.2019.12.008

Fu, S., Gu, H., & Yang, B. (2020). The Affordances of AI-Enabled Automatic Scoring Applications on Learners' Continuous Learning Intention: An Empirical Study in China. *British Journal of Educational Technology, 51*(5), 1674–1692. DOI: 10.1111/bjet.12995

Gallego, S., Rodriguez, R., & García, M. (2018). Design of a Conceptual Model for Improving Company Performance Based on Lean Management Applying the Viable System Model (VSM). *European Journal of Engineering and Formal Sciences, 2*(2), 49–63. DOI: 10.26417/ejef.v2i2.p50-64

Galperin, H., & Arcidiacono, M. (2021). Employment and the gender digital divide in Latin America: A decomposition analysis. *Telecommunications Policy, 45*(7), 102166. DOI: 10.1016/j.telpol.2021.102166

Gamage, P., & Halpin, E. F. (2007). E-Sri Lanka: Bridging the digital divide. *The Electronic Library, 25*(6), 693–710. DOI: 10.1108/02640470710837128

Gbadamosi, A. (2019). Marketing ethics, green and sustainable marketing. *Contemporary Issues in Marketing: Principles and Practice*, 185-213.

Geherson, C., Santi, P., & Ratti, C. (2021). Cybernetic cities: designing and controlling adaptive and robust urban systems. In Portugali, J. (Ed.), *Handbook on Cities and Complexity* (pp. 195–208). Edward Elgar Publishing., DOI: 10.4337/9781789900125.00018

GENÇ, N. (2023). Artificial Intelligence in Physical Education and Sports: New Horizons with ChatGPT. *Akdeniz Spor Bilimleri Dergisi*. DOI: 10.38021/asbid.1291604

Goel, A., Masurkar, S., & Pathade, G. R. (2024). An Overview of Digital Transformation and Environmental Sustainability: Threats, Opportunities, and Solutions. *Sustainability 2024, Vol. 16, Page 11079, 16*(24), 11079. DOI: 10.3390/su162411079

González Pascual, A., & García Jiménez, A. (2013). Internet y África: de la brecha a la esperanza digital. Redes, libertades y comunicación. http://hdl.handle.net/10115/15312

Goodfellow, I., Bengio, Y., & Courville, A. (2016). *Deep Learning*. MIT Press.

Goodman, J., & Salleh, A. (2013). The 'green economy': Class hegemony and counter-hegemony. *Globalizations*, *10*(3), 411–424. DOI: 10.1080/14747731.2013.787770

Goodwin, H. (2011). *Taking responsibility for tourism* (pp. 1-256). Woodeaton: Goodfellow Publishers Limited.

Gorur, R., Hoon, L., & Kowal, E. (2020). Computer Science Ethics Education in Australia - A Work in Progress. Paper presented at the 945–947. https://doi.org/ DOI: 10.1109/TALE48869.2020.9368375

Gössling, S., & Buckley, R. (2016). Carbon labels in tourism: Persuasive communication? *Journal of Cleaner Production*, *111*, 358–369. DOI: 10.1016/j.jclepro.2014.08.067

Gössling, S., & Buckley, R. (2016). *Carbon management in tourism: Mitigating the impacts on climate change*. Routledge.

Gössling, S., & Michael Hall, C. (2019). Sharing versus collaborative economy: How to align ICT developments and the SDGs in tourism? *Journal of Sustainable Tourism*, *27*(1), 74–96. DOI: 10.1080/09669582.2018.1560455

Gössling, S., & Peeters, P. (2015). Assessing tourism's global environmental impact 1900–2050. *Journal of Sustainable Tourism*, *23*(5), 639–659. DOI: 10.1080/09669582.2015.1008500

Gouseti, A., James, F., Fallin, L., & Burden, K. (2025). The Ethics of Using AI in K-12 Education: A Systematic Literature Review. *Technology, Pedagogy and Education*, *34*(2), 161–182. DOI: 10.1080/1475939X.2024.2428601

Graesser, A. C., Hu, X., Nye, B. D., VanLehn, K., Kumar, R., Heffernan, C., Heffernan, N., Woolf, B., Olney, A. M., Rus, V., Andrasik, F., Pavlik, P., Cai, Z., Wetzel, J., Morgan, B., Hampton, A. J., Lippert, A. M., Wang, L., Cheng, Q., & Baer, W. (2018). ElectronixTutor: An Intelligent Tutoring System with Multiple Learning Resources for Electronics. *International Journal of STEM Education*, *5*(1), 15–21. DOI: 10.1186/s40594-018-0110-y

Grassini, S. (2023). Shaping the Future of Education: Exploring the Potential and Consequences of AI and ChatGPT in Educational Settings. *Education Sciences*, *13*(7), 692. DOI: 10.3390/educsci13070692

Gray, S., Treacy, J., & Hall, E. T. (2019). Re-engaging disengaged pupils in physical education: An appreciative inquiry perspective. *Sport Education and Society*, *24*(3), 241–255. DOI: 10.1080/13573322.2017.1374942

*Green Erasmus Project*. (n.d.). Retrieved April 13, 2025, from https://project.greenerasmus.org/

Greenpeace. (2022). *Hospitality's carbon cover-up: How hotels mislead consumers on sustainability*. https://greenpeace.org/reports

Green, T. D., & Donovan, L. C. (2018). *Learning Anytime, anywhere through Technology Reconsidering Teaching and Learning for the iMaker Generation* (Hall, G. E., Quinn, L. F., & Gollnick, D. M., Eds.)., DOI: 10.1002/9781118955901.ch9

Grover, V., & Nandal, M. (2024). Education system using cloud computing: A proposed model. *Educational Perspectives on Digital Technologies in Modeling and Management*, 178–194. DOI: 10.4018/979-8-3693-2314-4.ch009

Guardeño Juan, M. (2025). Evolution concept of education for sustainable development and its integration into the curriculum of Spain and France. *European Public & Social Innovation Review*, *10*, 1–19. DOI: 10.31637/epsir-2024-1335

Guardeño Juan, M., & Monsalve Lorente, L. (2024). Education for sustainable development in the curriculum of Spain and Ireland. *Universidad y Sociedad*, *16*(1), 30–44.

Guelmami, N., Fekih-Romdhane, F., Mechraoui, O., & Bragazzi, N. L. (2023). Injury Prevention, Optimized Training and Rehabilitation: How Is AI Reshaping the Field of Sports Medicine. *New Asian Journal of Medicine*, *30–34*, 30–34. Advance online publication. DOI: 10.61838/kman.najm.1.1.6

Guerra, E., & Morris, E. A. (2018). Cities, transportation, and the future of work: Insights from urban science. *Transport Policy*, *69*, 149–157.

Gunawan, K. D. H., Liliasari, L., Kaniawati, I., & Setiawan, W. (2021). Implementation of Competency Enhancement Program for Science Teachers Assisted by Artificial Intelligence in Designing HOTS-Based Integrated Science Learning. *Jurnal Penelitian Dan Pembelajaran IPA*, *7*(1), 55–65. DOI: 10.30870/jppi.v7i1.8655

Gungor, V. C., Sahin, D., Kocak, T., Ergut, S., Buccella, C., Cecati, C., & Hancke, G. P. (2011). Smart grid technologies: Communication technologies and standards. *IEEE Transactions on Industrial Informatics*, *7*(4), 529–539. DOI: 10.1109/TII.2011.2166794

Guo, R. (2023). Analysis of Artificial Intelligence Technology and Its Application in Improving the Effectiveness of Physical Education Teaching. *International Journal of Web-Based Learning and Teaching Technologies, 19*(1), 1–15. DOI: 10.4018/IJWLTT.335115

Gupta, K. P., & Bhaskar, P. (2020). Inhibiting and Motivating Factors Influencing Teachers' Adoption of AI-Based Teaching and Learning Solutions: Prioritization Using Analytic Hierarchy Process. *Journal of Information Technology Education, 19*, 693–723. DOI: 10.28945/4640

Gupta, S., Shah, S., & Bansal, S. (2021). A comprehensive review on electric vehicle market development, battery technologies, and challenges. *Renewable & Sustainable Energy Reviews, 147*, 111221.

Hammes, F., Hagg, A., Asteroth, A., & Link, D. (2022). Artificial Intelligence in Elite Sports—A Narrative Review of Success Stories and Challenges. *Frontiers in Sports and Active Living, 4*, 861466. DOI: 10.3389/fspor.2022.861466 PMID: 35899138

Haraty, R. A., & Bitar, G. (2019). Associating learning technology to sustain the environment through green mobile applications. *Heliyon, 5*(1), e01141. Advance online publication. DOI: 10.1016/j.heliyon.2019.e01141 PMID: 30705989

Harris, J., Sims, R., & Xu, W. (2023). The future of media consumption in autonomous vehicles. *Digital Media Review, 28*(1), 45–60.

Hasanova, G. I., & Mammadova, Z. I. (2024). GREEN SCHOOLS: A BRIEF REVIEW. *GREEN ECONOMICS, 2*(2), 178–183. DOI: 10.62476/ge22.178

Hashem, I. A. T., Chang, V., Anuar, N. B., Adewole, K., Yaqoob, I., Gani, A., Ahmed, E., & Chiroma, H. (2016). The role of big data in smart city. *International Journal of Information Management, 36*(5), 748–758. DOI: 10.1016/j.ijinfomgt.2016.05.002

Hasselwander, M., Lütjens, K., Scheier, B., Frieske, B., Ennen, D., Scheelhaase, J., Kumar, S., & Maertens, S. (2025). Digital platforms, multimodal transport, and super apps. *Transportation Research Procedia, 86*, 207–212. DOI: 10.1016/j.trpro.2025.04.027

Helsper, E. (2021). The digital disconnect: The social causes and consequences of digital inequalities.

Herrera, P., Huepe, M., & Trucco, D. (2025). Education and Development of Digital Skills in Latin America and the Caribbean. Economic Commission for Latin America and the Caribbean (ECLAC). https://repositorio.cepal.org/bitstreams/1bcc9786-a37c-4325-ba30-efe8b5f26022/download

Holmes, W., Bektik, D., Whitelock, D., & Woolf, B. (2018). Ethics in AIED: Who Cares? Paper presented at the *19th International Conference on Artificial Intelligence in Education,10948* 551–553.

Holstein, K., & Doroudi, S. (2021). Equity and artificial intelligence in education: will" aied" amplify or alleviate inequities in education? arXiv preprint arXiv:2104.12920. https://doi.org//arXiv.2104.12920.DOI: 10.48550

Honeker, A. (2018). Smart Cities and Youth Unemployment: Mobile Technology as a Tool for Connecting Labor Supply and Demand. https://repositorio.utdt.edu/handle/20.500.13098/12257

Honey, M. (2008). Setting standards: certification programmes for ecotourism and sustainable tourism. In *Ecotourism and Conservation in the Americas* (pp. 234–261). CABI. DOI: 10.1079/9781845934002.0234

Hosford, K., Pitman, B., & Winter, M. (2021). Facilitators and barriers to the implementation of community-based transportation services for older adults: Evidence from six case studies. *Transportation Research Interdisciplinary Perspectives, 24*(101062), 101062. Advance online publication. DOI: 10.1016/j.trip.2024.101062

Huang, J., Shen, G., & Ren, X. (2021). Connotation Analysis and Paradigm shift of Teaching Design Under Artificial Intelligence Technology. *International Journal of Emerging Technologies in Learning (iJET), 16*(5), 73–86. https://doi.org/ . v16i05.20287DOI: 10.3991/ijet

Huang, J., & Chen, Z. (2016). The Research and Design of Web-Based Intelligent Tutoring System. *International Journal of Multimedia and Ubiquitous Engineering, 11*(6), 337–348. DOI: 10.14257/ijmue.2016.11.6.30

Huang, Z., Wang, W., Jia, Z., Wang, Z., Huang, Z., Wang, W., Jia, Z., & Wang, Z. (2024). Exploring the Integration of Artificial Intelligence in Sports Coaching: Enhancing Training Efficiency, Injury Prevention, and Overcoming Implementation Barriers. *Journal of Computer and Communications, 12*(12), 201–217. DOI: 10.4236/jcc.2024.1212012

Huertas Alcalá, S. (2021). *SOStenibilidad, una propuesta didáctica para Educación Primaria.* Ministry of Education and Vocational Training.

Huhn, S., Axt, M., Gunga, H. C., Maggioni, M. A., Munga, S., Obor, D., Sié, A., Boudo, V., Bunker, A., Sauerborn, R., Bärnighausen, T., & Barteit, S. (2022). The Impact of Wearable Technologies in Health Research: Scoping Review. *JMIR mHealth and uHealth, 10*(1), e34384. Advance online publication. DOI: 10.2196/34384 PMID: 35076409

Hu, Z., Liu, Z., & Su, Y. (2024a). AI-Driven Smart Transformation in Physical Education: Current Trends and Future Research Directions. *Applied Sciences (Basel, Switzerland), 14*(22), 10616. DOI: 10.3390/app142210616

Idrees, A. K., & Jawad, L. W. (2023). Energy-efficient Data Processing Protocol in edge-based IoT networks. *Annales des Télécommunications, 78*(5–6), 347–362. DOI: 10.1007/s12243-023-00957-8

IEA. (2020). *Smart Grids: Tracking Report*. International Energy Agency. https://www.iea.org

IEA. (2023). *Global EV Outlook 2023*. International Energy Agency. https://www.iea.org/reports/global-ev-outlook-2023

International Energy Agency (IEA). (2021). *Global EV Outlook 2021: Accelerating the transition to electric mobility*. IEA Publications.

International Organization for Migration. (2025). IOM - UN migration. Retrieved March 13, 2025, from https://www.iom.int/

Jackson, M. C. (2000). *Systems Approaches to Management*. Springer US.

Jackson, M., & Cossitt, B. (2015). *Is Intelligent Online Tutoring Software Useful in Refreshing Financial Accounting Knowledge? Advances in Accounting Education: Teaching and Curriculum Innovations*. Emerald Group Publishing Limited., DOI: 10.1108/S1085-462220150000016001

Ji, T., Luo, C., Yu, L., Wang, Q., Chen, S., Thapa, A., & Li, P. (2022). Energy-Efficient computation offloading in mobile edge computing systems with uncertainties. *IEEE Transactions on Wireless Communications, 21*(8), 5717–5729. DOI: 10.1109/TWC.2022.3142685

Johnson, A. M., Zhou, C., Haviland, M., & Mendoza, J. A. (2024). Evaluation of a walking school bus program: A cluster randomized controlled trial. *The International Journal of Behavioral Nutrition and Physical Activity, 21*(55), 55. Advance online publication. DOI: 10.1186/s12966-024-01602-w PMID: 38730407

Jones, M., & Patel, R. (2023). Consumer skepticism and the greenwashing dilemma. *International Journal of Consumer Studies, 47*(1), 89–102. DOI: 10.5678/ijcs.47189

Jones, M., Thompson, E., & Ruiz, L. (2023). Transparent sustainability: The new frontier in eco-marketing. *Journal of Environmental Communication, 9*(3), 33–50. DOI: 10.3210/jec.v9i3.987

Kabudi, T., Pappas, I., & Olsen, D. H. (2021). AI-Enabled Adaptive Learning Systems: A Systematic Mapping of the Literature. *Computers and Education. Artificial Intelligence*, *2*, 100017. DOI: 10.1016/j.caeai.2021.100017

Kahneman, D., & Tversky, A. (1979). Prospect theory: An analysis of decision under risk. *Econometrica*, *47*(2), 263–292. DOI: 10.2307/1914185

Kapoor, A., Sharma, S., & Gupta, R. (2022). Sustainability in advertising: The role of eco-friendly brands in a green economy. *Journal of Sustainable Marketing*, *18*(4), 223–240.

Karaca, O., Çalışkan, S. A., & Demir, K. (2021). Medical Artificial Intelligence Readiness Scale for Medical Students (MAIRS-MS) – Development, Validity and Reliability Study. *BMC Medical Education*, *21*(1), 10. DOI: 10.1186/s12909-021-02546-6

Kasneci, E., Sessler, K., Kuechemann, S., Bannert, M., Dementieva, D., Fischer, F., Gasser, U., Groh, G., Guennemann, S., Huellermeier, E., Krusche, S., Kutyniok, G., Michaeli, T., Nerdel, C., Pfeffer, J., Poquet, O., Sailer, M., Schmidt, A., Seidel, T., & Kasneci, G. (2023). ChatGPT for Good? On Opportunities and Challenges of Large Language Models for Education. *Learning and Individual Differences*, *103*, 102274. DOI: 10.1016/j.lindif.2023.102274

Kaya, F. (2025). The Use of Artificial Intelligence in Physical Education and Movement Development in Children. *Spor Eğitim Dergisi*, *9*(1), 1–13. DOI: 10.55238/seder.1583190

Khan, M. K., Farouk, A., Yu, H., & Mi, Y. (2023). Application Model for Innovative Sports Practice Teaching in Colleges Using Internet of Things and Artificial Intelligence. *Electronics 2023, Vol. 12, Page 874*, *12*(4), 874. DOI: 10.3390/electronics12040874

Khanh, Q. V., Nguyen, V., Minh, Q. N., Van, A. D., Anh, N. L., & Chehri, A. (2023). An efficient edge computing management mechanism for sustainable smart cities. *Sustainable Computing: Informatics and Systems*, *38*, 100867. DOI: 10.1016/j.suscom.2023.100867

Khan, S., Shah, I. A., Aurangzeb, K., Ahmad, S., Khan, J. A., & Anwar, M. S. (2024). Energy efficient task scheduling using fault tolerance technique for IoT applications in FOG computing environment. *IEEE Internet of Things Journal*, *11*(24), 39009–39019. DOI: 10.1109/JIOT.2024.3403003

Kharbat, F. F., Alshawabkeh, A., & Woolsey, M. L. (2021). Identifying Gaps in Using Artificial Intelligence to Support Students with Intellectual Disabilities from Education and Health Perspectives. *Aslib Journal of Information Management, 73*(1), 101–128. DOI: 10.1108/AJIM-02-2020-0054

Kilbourne, W. E. (2004). Sustainable marketing: The new marketing paradigm for a new millennium. *Marketing Intelligence & Planning, 22*(7), 702–710.

Kim, M., Lee, J., & Lee, H. (2019). Smart cities and urban policy in South Korea. *Cities (London, England), 89*, 48–56.

Kim, M., Lee, J., & Lee, H. (2019). Sustainable urban infrastructure through smart waste management. *Journal of Cleaner Production, 220*, 123–134.

Kitchin, R. (2016). The ethics of smart cities and urban science. *Philosophical Transactions. Series A, Mathematical, Physical, and Engineering Sciences, 374*(2083), 20160115. DOI: 10.1098/rsta.2016.0115 PMID: 28336794

Koh, J. K. (2024). Analyzing PE Teachers' Interest and Execution in Using AI in Education. *Journal of Research in Curriculum Instruction, 28*(1), 44–54.

Kollmann, T., Henseler, J., & Buxmann, P. (2020). Digitalization and its impact on consumer behavior in the automotive sector: A study of electric vehicle drivers. *The Journal of Consumer Research, 47*(2), 219–235.

Konecny, J., Prauzek, M., & Borova, M. (2023). Fuzzy controlled Wavelet-Based edge Computing method for Energy-Harvesting IoT sensors. *IEEE Internet of Things Journal, 10*(21), 18909–18918. DOI: 10.1109/JIOT.2023.3292915

Kong, S., & Yang, Y. (2024). A Human-Centred Learning and Teaching Framework Using Generative Artificial Intelligence for Self-Regulated Learning Development Through Domain Knowledge Learning in K–12 Settings. *IEEE Transactions on Learning Technologies, 17*, 1562–1573. DOI: 10.1109/TLT.2024.3392830

Kong, X., Duan, G., Hou, M., Shen, G., Wang, H., Yan, X., & Collotta, M. (2022). Deep reinforcement Learning-Based Energy-Efficient edge computing for internet of vehicles. *IEEE Transactions on Industrial Informatics, 18*(9), 6308–6316. DOI: 10.1109/TII.2022.3155162

Kos, A., & Umek, A. (2018). *Biomechanical Biofeedback Systems and Applications*. DOI: 10.1007/978-3-319-91349-0

Kotler, P., & Keller, K. L. (2016). *Marketing management* (15th ed.). Pearson.

Krishna Azith Teja Ganti, V., Pandugula, C., Naga Subhash Polineni, T., & Mallesham, G. (n.d.). Transforming Sports Medicine with Deep Learning and Generative AI. *Personalized Rehabilitation Protocols and Injury Prevention Strategies for Professional Athletes.*, *22*(1), 2023–3868.

Latif, E., Zhai, X., & Liu, L. (2025). *AI Gender Bias, Disparities, and Fairness: Does Training Data Matter?* Lee, H. S., & Lee, J. (2021). Applying Artificial Intelligence in Physical Education and Future Perspectives. *Sustainability (Basel)*, *13*(1), 351. DOI: 10.3390/su13010351

Laufer, W. S. (2003). Social accountability and corporate greenwashing. *Journal of Business Ethics*, *43*(3), 253–261. DOI: 10.1023/A:1022962719299

Leape, J. (2006). The London congestion charge. *The Journal of Economic Perspectives*, *20*(4), 157–176. DOI: 10.1257/jep.20.4.157

Lee, J., Park, J., & Kim, H. (2018). Structural health monitoring for urban bridge safety in Seoul using sensor technologies. *Sensors (Basel)*, *18*(3), 845. PMID: 29533994

Lee, S.-B. (2024). The Effect of AI-based Physical Education Convergence Program on Convergent Thinking and Problem-solving Skills. *Journal of the Korean Society for Wellness*, *19*(4), 295–300. DOI: 10.21097/ksw.2024.11.19.4.295

Lee, Y., & Kim, D. (2025). Evolving Professionalism in the AI Era: Implementing Generative AI in Physical Education. *Journal of Physical Education, Recreation & Dance*, *96*(3), 8–13. DOI: 10.1080/07303084.2024.2444586

Leonidou, C. N., & Skarmeas, D. (2017). Gray shades of green: Causes and consequences of green skepticism. *Journal of Business Ethics*, *144*(2), 401–415. DOI: 10.1007/s10551-015-2829-4

Leonidou, L. C., (2020). "Green" advertising effects on consumer responses: A contingency perspective. *Journal of Advertising*, *49*(1), 1–20.

Li, D., Lan, M., & Hu, Y. (2022). Energy-saving service management technology of internet of things using edge computing and deep learning. *Complex & Intelligent Systems*, *8*(5), 3867–3879. DOI: 10.1007/s40747-022-00666-0

Lim, M. (2017). Freedom to hate: Social media, algorithmic enclaves, and the rise of tribal nationalism in Indonesia. *Critical Asian Studies*, *49*(3), 411–427. DOI: 10.1080/14672715.2017.1341188

Liu, L., Pu, Y., Fan, J., Yan, Y., Liu, W., Luo, K., Wang, Y., Zhao, G., Chen, T., Puiu, P. D., & Huang, H. (2024). Wearable Sensors, Data Processing, and Artificial Intelligence in Pregnancy Monitoring: A Review. *Sensors 2024, Vol. 24, Page 6426, 24*(19), 6426. DOI: 10.3390/s24196426

Liu, D., Liang, H., Zeng, X., Zhang, Q., Zhang, Z., & Li, M. (2022). Edge computing application, architecture, and challenges in ubiquitous power internet of things. *Frontiers in Energy Research, 10*, 850252. Advance online publication. DOI: 10.3389/fenrg.2022.850252

Liu, J., Li, C., & Luo, Y. (2024). Efficient resource allocation for IoT applications in mobile edge computing via dynamic request scheduling optimization. *Expert Systems with Applications, 255*, 124716. DOI: 10.1016/j.eswa.2024.124716

Liu, S., Chen, S., & Wang, X. (2020). Hyper-targeted advertising: The role of autonomous vehicles in personalized marketing. *Journal of Advertising Research, 62*(3), 187–200.

Liu, Z., Ren, Y., Kong, X., & Liu, S. (2022). Learning Analytics Based on Wearable Devices: A Systematic Literature Review From 2011 to 2021. *Journal of Educational Computing Research, 60*(6), 1514–1557. DOI: 10.1177/07356331211064780

Loján Cueva, A. B. (2023). *Artificial Intelligence to diagnose low back pain using motion captures*. https://repositorio.yachaytech.edu.ec/handle/123456789/667

Long, D., & Magerko, B. (2020). What is AI Literacy? Competencies and Design Considerations. Paper presented at the *Proceedings of the 2020 CHI Conference on Human Factors in Computing Systems,* 1–16. https://doi.org/DOI: 10.1145/3313831.3376727

Longo, F., & Vetrano, A. (2021). Smart Waste Management in Smart Cities: A Decision Support Approach. *Simulation Modelling Practice and Theory, 108*, 102275.

López López, H. L., Rivera Escalera, A., & Cruz García, C. R. (2023). PERSONALIZACIÓN DEL APRENDIZAJE CON INTELIGENCIA ARTIFICIAL EN LA EDUCACIÓN SUPERIOR. *Revista Digital de Tecnologías Informáticas y Sistemas, 7*(1), 123–128. DOI: 10.61530/redtis.vol7.n1.2023.165.123-128

Lucas, R. E. (1990). VWhy Doesnpt Capital Flow from Rich to Poor Countries? V. V American Economic Review, 80(2), 92q96.

Luckin, R., & Holmes, W. (2016). Intelligence unleashed: An argument for AI in education.

Luo, Y., Luo, L., & Luo, D. (2024). Research on the Enhancement Path of Physical Education Teaching in Colleges and Universities under Artificial Intelligence Technology. *Applied Mathematics and Nonlinear Sciences*, *9*(1), 20241770. Advance online publication. DOI: 10.2478/amns-2024-1770

Lyon, T. P., & Montgomery, A. W. (2015). The means and end of greenwash. *Organization & Environment*, *28*(2), 223–249. DOI: 10.1177/1086026615575332

Mahenge, M. P. J., Li, C., & Sanga, C. A. (2022). Energy-efficient task offloading strategy in mobile edge computing for resource-intensive mobile applications. *Digital Communications and Networks*, *8*(6), 1048–1058. DOI: 10.1016/j.dcan.2022.04.001

Mao, Y., You, C., Zhang, J., Huang, K., & Letaief, K. B. (2017b). A survey on Mobile Edge Computing: The Communication Perspective. *IEEE Communications Surveys and Tutorials*, *19*(4), 2322–2358. DOI: 10.1109/COMST.2017.2745201

Martín-García, J., Dies Álvarez, M. E., & Afonso, A. S. (2024). Understanding Science Teachers' Integration of Active Methodologies in Club Settings: An Exploratory Study. *Education Sciences*, *14*(1), 106. Advance online publication. DOI: 10.3390/educsci14010106

Martín-Rodríguez, A., & Madrigal-Cerezo, R. (2025). Technology-Enhanced Pedagogy in Physical Education: Bridging Engagement, Learning, and Lifelong Activity. *Education Sciences*, *15*(4), 409. DOI: 10.3390/educsci15040409

Marwala, T., & Mpedi, L. G. (2024). Social Protection. In Artificial Intelligence and the Law (pp. 119-141). Singapore: Springer Nature Singapore. https://doi.org/ DOI: 10.1007/978-981-97-2827-5_7

McKinsey & Company. (2021). *The future of mobility: Global automotive study*. McKinsey & Company.

Meadows, D. H., Meadows, D. L., Randers, J., & Behrens, W. W.III. (1972). *The limits to growth*. Universe Books.

Medina, E. (2011). *Cybernetic Revolutionaries: Technology and Politics in Allende's Chile*. MIT Press. DOI: 10.7551/mitpress/8417.001.0001

Memarian, B., & Doleck, T. (2023). ChatGPT in Education: Methods, Potentials, and Limitations. *Computers in Human Behavior: Artificial Humans*, *1*(2), 100022. DOI: 10.1016/j.chbah.2023.100022

Memarian, B., & Doleck, T. (2023). Fairness, Accountability, Transparency, and Ethics (FATE) in Artificial Intelligence (AI) and higher education: A systematic review. *Computers and Education: Artificial Intelligence*, *5*, 100152. DOI: 10.1016/j.caeai.2023.100152

Meyer, K. E., Li, J., Brouthers, K. D., & Jean, R. J. (2023). International business in the digital age: Global strategies in a world of national institutions. *Journal of International Business Studies*, *54*(4), 577–598. DOI: 10.1057/s41267-023-00618-x PMID: 37305175

Miao, F., Holmes, W., Huang, R., & Zhang, H. (2021). *AI and Education: A Guidance for Policymakers*. Unesco Publishing.

Migration Policy Institute (MPI). (2025). Migration Policy Institute. https://www.migrationpolicy.org/

Milani, S., Garlisi, D., Carugno, C., Tedesco, C., & Chatzigiannakis, I. (2024). Edge2LoRa: Enabling edge computing on long-range wide-area Internet of Things. *Internet of Things : Engineering Cyber Physical Human Systems*, *27*, 101266. DOI: 10.1016/j.iot.2024.101266

Millar, M., & Baloglu, S. (2011). Hotel guests' preferences for green guest room attributes. *Cornell Hospitality Quarterly*, *52*(3), 302–311. DOI: 10.1177/1938965511409031

Miller, L., Thompson, S., & Wallis, D. (2021). Ethical considerations in data-driven advertising: Privacy and bias in autonomous vehicle marketing. *Journal of Business Ethics*, *162*(1), 73–89.

Minh, Q. N., Nguyen, V., Quy, V. K., Ngoc, L. A., Chehri, A., & Jeon, G. (2022). Edge computing for IoT-Enabled smart Grid: The future of energy. *Energies*, *15*(17), 6140. DOI: 10.3390/en15176140

Ministry for Ecological Transition and the Demographic Challenge (n.d.). *Educational Centres Towards Sustainability*.

Ministry of Housing and Urban Affairs, Government of India. (2022). *Smart Cities Mission Annual Report*. Retrieved from https://smartcities.gov.in

Mohammed, P. S., & Watson, E. '. (2019). Towards Inclusive Education in the Age of Artificial Intelligence: Perspectives, Challenges, and Opportunities. *Artificial Intelligence and Inclusive Education: Speculative Futures and Emerging Practices*, 17–37. https://doi.org/DOI: 10.1007/978-981-13-8161-4_2

Mohebbi, A. (2025). Enabling Learner Independence and Self-Regulation in Language Education Using AI Tools: A Systematic Review. *Cogent Education*, *12*(1), 2433814. DOI: 10.1080/2331186X.2024.2433814

Mokmin, N. A. M. (2020). The Effectiveness of a Personalized Virtual Fitness Trainer in Teaching Physical Education by Applying the Artificial Intelligent Algorithm. *International Journal of Human Movement and Sports Sciences*, *8*(5), 258–264. DOI: 10.13189/saj.2020.080514

Monge Vera, M. M., Villamagua Jiménez, G. M., Aroca Izurieta, C. E., Chico Guzmán, B. A., & López Velasco, J. E. (2024). Personalización del proceso de aprendizaje mediante inteligencia artificial. *LATAM Revista Latinoamericana de Ciencias Sociales y Humanidades*, *5*(3). Advance online publication. DOI: 10.56712/latam.v5i3.2076

Montiel-Ruiz, F. J., Calderón, A., Sánchez-Vera, M. del M., & Solano-Fernández, I. M. (2024). Redes sociales y aprendizaje móvil en Educación Física: Creación colaborativa de principios de diseño. *Edutec. Revista Electrónica de Tecnología Educativa*, *89*(89), 190–209. DOI: 10.21556/edutec.2024.89.3191

Moore, R., Vitale, D., & Stawinoga, N. (2018). The digital divide and educational equity. *Insights in Education and Work, 1*–10.

Moorhouse, B. L. (2024). Beginning and First-Year Language Teachers' Readiness for the Generative AI Age. *Computers and Education: Artificial Intelligence*, *610*, 100201. Advance online publication. DOI: 10.1016/j.caeai.2024.100201

Moran, M., & King, M. (2021). Augmented reality in advertising: Opportunities in autonomous vehicles. *Journal of Interactive Marketing*, *39*, 50–61.

Moreira, J. L. C., Aguilera, A. G. F., Solís, I. A. M., Dután, J. F. L., Centeno, M. V. C., & Taco, N. B. I. (2024). Reducción de la brecha digital en zonas rurales: soluciones tecnológicas para una educación

Morsing, M., & Schultz, M. (2006). Corporate social responsibility communication: Stakeholder information, response and involvement strategies. *Business Ethics (Oxford, England)*, *15*(4), 323–338. DOI: 10.1111/j.1467-8608.2006.00460.x

Morsing, M., Schultz, M., & Nielsen, K. U. (2008). The 'Catch 22'of communicating CSR: Findings from a Danish study. *Journal of Marketing Communications*, *14*(2), 97–111. DOI: 10.1080/13527260701856608

Mou, S. (2025). The integration of mobile technology in biomechanics education: Advancing knowledge and practice. *Molecular & Cellular Biomechanics*, *22*(3), 1410. DOI: 10.62617/mcb1410

Mouta, A., Torrecilla-Sanchez, E., & Pinto-Llorente, A. (2024). Design of a Future Scenarios Toolkit for an Ethical Implementation of Artificial Intelligence in Education. *Education and Information Technologies*, *29*(9), 10473–10498. DOI: 10.1007/s10639-023-12229-y

Murphy, S., & Regenold, T. (2023). Applying the Viable System Model to Local Government in a Post-Pandemic Context. In Azukas, M. E., & Kim, M. (Eds.), *Reimagining Systems Thinking in a Post-pandemic World* (pp. 79–96). IGI Global., DOI: 10.4018/978-1-6684-7285-9.ch004

Nam, T. (2011). Conceptualizing smart city with dimensions of technology, people, and institutions. *Proceedings of the 12th Annual International Conference on Digital Government Research*. DOI: 10.1145/2037556.2037602

National Highway Traffic Safety Administration (NHTSA). (2021). *Critical reasons for crashes investigated in the National Motor Vehicle Crash Causation Survey*. NHTSA.

Navarro-González, I., & Gavari-Starkie, E. (2024). Sustainable and resilient education in the LOMLOE curriculum. *Revista Española de Educación Comparada*, (45), 376–397. DOI: 10.5944/reec.45.2024.37968

Neirotti, P., De Marco, A., Cagliano, A. C., Mangano, G., & Scorrano, F. (2014). Current trends in Smart City initiatives: Some stylised facts. *Cities (London, England)*, *38*, 25–36. DOI: 10.1016/j.cities.2013.12.010

Ng, D. T. K., Lee, M., Tan, R. J. Y., Hu, X., Downie, J. S., & Chu, S. K. W. (2023). A Review of AI Teaching and Learning from 2000 to 2020. *Education and Information Technologies*, *28*(7), 8445–8501. DOI: 10.1007/s10639-022-11491-w

Nielsen. (2021). *The connected consumer in the age of autonomous vehicles: Implications for marketers*. Nielsen Insights.

Novo, M. (1998). La educación ambiental: bases éticas, conceptuales y metodologías.

Ntoutsi, E., Fafalios, P., Gadiraju, U., Iosifidis, V., Nejdl, W., Vidal, M., Ruggieri, S., Turini, F., Papadopoulos, S., Krasanakis, E., Kompatsiaris, I., Kinder-Kurlanda, K., Wagner, C., Karimi, F., Fernandez, M., Alani, H., Berendt, B., Kruegel, T., Heinze, C., & Staab, S. (2020). Bias in Data-Driven Artificial Intelligence Systems-An Introductory Survey. *Wiley Interdisciplinary Reviews. Data Mining and Knowledge Discovery*, *10*(3), e1356. DOI: 10.1002/widm.1356

Núñez, M. S., Morales, M. P., & Montañés, A. S. (2018). *Cada Gota Cuenta. Actas de los Premios de Investigación e Innovación de Educación Secundaria Obligatoria*. Bachillerato y Formación Profesional de Castilla y León.

Nye, B. D. (2015). Intelligent Tutoring Systems by and for the Developing World: A Review of Trends and Approaches for Educational Technology in a Global Context. *International Journal of Artificial Intelligence in Education, 25*(2), 177–203. DOI: 10.1007/s40593-014-0028-6

Nyilasy, G., Gangadharbatla, H., & Paladino, A. (2014). Perceived greenwashing: The interactive effects of green advertising and corporate environmental performance on consumer reactions. *Journal of Business Ethics, 125*(4), 693–707. DOI: 10.1007/s10551-013-1944-3

Ofosu-Asare, Y. (2024). Developing classroom ICT teaching techniques, principles and practice for teachers in rural Ghana without access to computers or internet: A framework based on literature review. *The International Journal of Information and Learning Technology, 41*(3), 262–279. DOI: 10.1108/IJILT-04-2023-0045

Oladele, T., Oyenuga, M., Oyetunde, O. T., Oyedele, O. M., & Jaccinta, A. G. (2023). *Green Management Practices and Organisational Efficiency in a Higher Institution. Article in Budapest International Research and Critics in Linguistics and Education (BirLE).* Journal., DOI: 10.33258/birle.v6i3.7716

Olvera, E. Y. M., Bastidas, E. G. J., & Espinoza, G. J. M. (2024). Analysis of the Digital Divide and Access to Technological Resources in Secondary Education Institutions in Ecuador. *Ciencia Latina Multidisciplinary Scientific Journal, 8*(2), 6698–6719. DOI: 10.37811/cl_rcm.v8i2.11086

Ompoc, H. M., & Aguinaldo, J. (2024). The Impact of Artificial Intelligence (AI) Integration on Student Engagement and Learning Outcomes of Tertiary Physical Education Students. *Pedagogy Review: An International Journal of Educational Theories. Approaches and Strategies, 4*(1), 80–86. DOI: 10.62718/vmca.pr-ijetas.4.1.SC-1224-006

Oregi, I. L., Urra-Uriarte, S., Gonzalez, I., Anaya, M., & Molina-Costa, P. (2025). Enhancing Citizen Participation in Citizen-Centered Smart Cities: Insights from Two European Case Studies. *Urban Science (Basel, Switzerland), 9*(5), 140. DOI: 10.3390/urbansci9050140

Organisation for Economic Co-operation and Development (OECD). (2021). *Digital education outlook 2021: Pushing the frontiers with AI.* OECD Publishing., DOI: 10.1787/589b283f-

Ottman, J. A. (2011). *The new rules of green marketing: Strategies, tools, and inspiration for sustainable branding*. Greenleaf Publishing.

Ouyang, F., & Zhang, L. (2024). AI-Driven Learning Analytics Applications and Tools in Computer-Supported Collaborative Learning: A Systematic Review. *Educational Research Review*, *44*, 100616. DOI: 10.1016/j.edurev.2024.100616

Pardellas-Santiago, M., García-Fernández, J., Majadas-Andray, J., Fernández-Ramos, J., & Sampedro Ortega, Y. (2024). *Estudio comparativo del Plan de Acción de Educación Ambiental para la Sostenibilidad (PAEAS) con las Estrategias Autonómicas de Educación Ambiental*. CENEAM / OAPN.

Parguel, B., Benoît-Moreau, F., & Larceneux, F. (2011). How sustainability ratings might deter 'greenwashing': A closer look at ethical corporate communication. *Journal of Business Ethics*, *102*(1), 15–28. DOI: 10.1007/s10551-011-0901-2

Peattie, K. (2010). Green marketing: Legend, myth, delusion, or panacea? *Journal of Strategic Marketing*, *18*(2), 109–117.

Peattie, K., & Crane, A. (2005). Green marketing: Legend, myth, farce or prophesy? *Qualitative Market Research*, *8*(4), 357–370. DOI: 10.1108/13522750510619733

Pecan Street Inc. (2019). *Data-Driven Energy Solutions*. https://www.pecanstreet.org

Pedregosa, F., Varoquaux, G., & Gramfort, A. (2011). Scikit-learn: Machine learning in Python. *Journal of Machine Learning Research*, *12*, 2825–2830.

Pereira, R. (2003). Libro blanco de la Educación Ambiental en España. *Ministry of the Environment. Secretaría General de Medio Ambiente.*

Pesovski, I., Santos, R., Henriques, R., & Trajkovik, V. (2024). Generative AI for Customizable Learning Experiences. *Sustainability (Basel)*, *16*(7), 3034. DOI: 10.3390/su16073034

Pierce, G., & Shoup, D. (2013). Getting the prices right: An evaluation of pricing parking by demand in San Francisco. *Journal of the American Planning Association*, *79*(1), 67–81. DOI: 10.1080/01944363.2013.787307

Piñeiro Méndez, A. (2022). Sustainable development goals in primary education and their implementation through information and communication technologies.

Pinillos, G., Masferrer, C., & Oropeza, N. (2019). Inequalities, Migration, and Borders: A Proposal for a Multidimensional Approach. *Notes on Migration and Inequalities*, *3*, 2–13.

Pira, M. (2021). A novel taxonomy of smart sustainable city indicators. *Humanities & Social Sciences Communications*, *8*(197), 197. Advance online publication. DOI: 10.1057/s41599-021-00879-7

Prakash, J., Swathiramya, R., Balambigai, G., Menaha, R., & Abhirami, J. S. (2024). AI-Driven Real-Time Feedback System for Enhanced Student Support: Leveraging Sentiment Analysis and Machine Learning Algorithms. *International Journal of Computational and Experimental Science and Engineering*, *10*(4), 1567–1574. DOI: 10.22399/ijcesen.780

Premalatha, B., & Prakasam, P. (2023). Optimal Energy-efficient Resource Allocation and Fault Tolerance scheme for task offloading in IoT-FoG Computing Networks. *Computer Networks*, *238*, 110080. DOI: 10.1016/j.comnet.2023.110080

Prieto Andreu, J. M. (2024). Paleocontacto GT: fusionando ejercicio físico con un juego de mesa (GT Paleocontact: fusing physical exercise with a board game). *Retos*, *59*, 892–902. DOI: 10.47197/retos.v59.107734

Prieto-Castillo, L., Mijarra, J. J., & Rubio-Romero, J. C. (2021). La publicidad verde como herramienta para el consumo responsable: Un análisis de la percepción del consumidor. *Revista Mediterránea de Comunicación*, *12*(1), 145–159.

Pueo, B., Penichet-Tomas, A., & Jimenez-Olmedo, J. M. (2020). Validity, reliability and usefulness of smartphone and kinovea motion analysis software for direct measurement of vertical jump height. *Physiology & Behavior*, *227*, 113144. Advance online publication. DOI: 10.1016/j.physbeh.2020.113144 PMID: 32822710

Pulido-Fernández, J. I., & López-Sánchez, Y. (2016). La cadena de valor del destino como herramienta innovadora para el análisis de la sostenibilidad de las políticas turísticas. El caso de España. *Innovar (Universidad Nacional de Colombia)*, *26*(59), 155–176. DOI: 10.15446/innovar.v26n59.54369

Quijano, A. (2004). The Labyrinth of Latin America: Are There Other Ways Out? *Journal of Iberian and Latin American Research*, *10*(2), 173–196. DOI: 10.1080/13260219.2004.10426802

Quy, N. M., Ngoc, L. A., Ban, N. T., Van Hau, N., & Quy, V. K. (2023). Edge Computing for Real-Time Internet of Things Applications: Future Internet Revolution. *Wireless Personal Communications*, *132*(2), 1423–1452. DOI: 10.1007/s11277-023-10669-w

Rabelo, A., Rodrigues, M. W., Nobre, C., Isotani, S., & Zárate, L. (2024). Educational Data Mining and Learning Analytics: A Review of Educational Management in e-learning. *Information Discovery and Delivery*, *52*(2), 149–163. DOI: 10.1108/IDD-10-2022-0099

Rahman, I., Park, J., & Chi, C. G. (2017). Consequences of "greenwashing": Consumers' reactions to hotels' green initiatives. *International Journal of Contemporary Hospitality Management, 27*(6), 1054–1081. DOI: 10.1108/IJCHM-04-2014-0202

Rahman, S. U., & Nguyen-Viet, B. (2023). Towards sustainable development: Coupling green marketing strategies and consumer perceptions in addressing greenwashing. *Business Strategy and the Environment, 32*(4), 2420–2433. DOI: 10.1002/bse.3256

Rainforest Alliance. (2022). *Tourism and sustainability: Certification gaps in global hotel chains*. https://www.rainforest-alliance.org/publications

Ratcliffe, J. H., Strang, S. J., Taylor, R. B., & Sorg, E. T. (2019). The impact of intelligence-led policing in a major U.S. city. *Justice Quarterly, 36*(5), 831–856.

Raymond, S. J., Cecchi, N. J., Alizadeh, H. V., Callan, A. A., Rice, E., Liu, Y., Zhou, Z., Zeineh, M., & Camarillo, D. B. (2021). Physics-informed machine learning improves detection of head impacts. *Annals of Biomedical Engineering, 50*(11), 1534–1545. DOI: 10.1007/s10439-022-02911-6 PMID: 35303171

Real-Pérez, M., Sánchez-Oliva, D., & Moledo, C. P. (2021). Proyecto África "La Leyenda de Faro": Efectos de una metodología basada en la gamificación sobre la motivación situacional respecto al contenido de expresión corporal en Educación Secundaria (Africa Project "La Leyenda de Faro": Effects of a methodology. *Retos, 42*, 567–574. DOI: 10.47197/retos.v42i0.86124

Restrepo, E. G. Y., Baldassarre, M., & Boticario, J. G. (2019). Accessibility, Biases and Ethics in Chatbots and Intelligent Agents for Education. Paper presented at the *EDULEARN19 Proceedings,* 8824–8833. https://doi.org/DOI: 10.21125/edulearn.2019.2196

Richmond, A. H. (1984). Adaptation and Sociocultural Conflicts in Immigrant-Receiving Countries. https://unesdoc.unesco.org/ark:/48223/pf0000062549_spa

Robinsha, S. D., & Amutha, B. (2023). IoT architecture for energy management in smart cities. *International Journal of Services Operations and Informatics, 12*(4), 325–343. DOI: 10.1504/IJSOI.2023.137474

Robinsha, S. D., & Amutha, B. (2024). Velocious: A Resilient IoT Architecture for 6G Based Intelligent Transportation System with Expeditious Movement Mechanism. *Wireless Personal Communications*. Advance online publication. DOI: 10.1007/s11277-024-11072-9

Romero, A. M. M. (2020). The Generational Digital Divide. *Labor Issues: Andalusian Journal of Work and Social Welfare*, (151), 77–93.

S. BELLEZA, S., OMPOC, R., & VESTIL, C. (2021). Challenges of PE Teachers to Educational Technologies in New Normal: The Changing Role of Teachers. *International Journal of Research Publications, 86*(1). Advance online publication. DOI: 10.47119/IJRP1008611020212320

Saadallah, N. R., & Alabady, S. A. (2024). A comprehensive study on energy-efficient-based routing protocols in the Internet of things Part I: Definition and classification. *Iran Journal of Computer Science, 7*(3), 631–661. DOI: 10.1007/s42044-024-00180-1

Saghiri, M. A., Vakhnovetsky, J., & Nadershahi, N. (2022). Scoping Review of Artificial Intelligence and Immersive Digital Tools in Dental Education. *Journal of Dental Education, 86*(6), 736–750. DOI: 10.1002/jdd.12856

Salas-Pilco, S. Z., Xiao, K., & Hu, X. (2022). Artificial Intelligence and Learning Analytics in Teacher Education: A Systematic Review. *Education Sciences, 12*(8), 569. DOI: 10.3390/educsci12080569

Saleem, M., Shakir, M., Usman, M., Bajwa, M., Shabbir, N., Ghahfarokhi, P. S., & Daniel, K. (2023). Integrating Smart Energy Management System with Internet of Things and Cloud Computing for Efficient Demand Side Management in Smart Grids. *Energies, 16*(12), 4835. DOI: 10.3390/en16124835

Satariano, A. (2018). Atlanta's Cyberattack and the Age of Digital Disaster. *The New York Times*. https://www.nytimes.com

Sauv, Ã, L. (2005). Currents in environmental education: Mapping a complex and evolving pedagogical field. [CJEE]. *Canadian Journal of Environmental Education*, •••, 11–37.

Scala, D., Aguilar Cuesta, Á. I., & Cañizares, M. C. (2024). Integrating smart cities in Primary Education: An approach towards sustainability and urban resilience. ENSAYOS [Web link: http://www.revista.uclm.es/index.php/ensayos]. *Journal of the Faculty of Education of Albacete, 39*(2), 127–144.

Scheyvens, R. (2011). The challenge of sustainable tourism development in the Maldives: Understanding the social and political dimensions of sustainability. *Asia Pacific Viewpoint, 52*(2), 148–164. DOI: 10.1111/j.1467-8373.2011.01447.x

Schiff, D. (2022). Education for AI, not AI for Education: The Role of Education and Ethics in National AI Policy Strategies. *International Journal of Artificial Intelligence in Education, 32*(3), 527–563. DOI: 10.1007/s40593-021-00270-2

Schubert, C. (n.d.). *Research Guides: Artificial Intelligence (AI) in Education: AI and Ethics*. Retrieved April 16, 2025, from https://guides.lib.jmu.edu/AI-in-education/ethics

Shajari, S., Kuruvinashetti, K., Komeili, A., & Sundararaj, U. (2023). The Emergence of AI-Based Wearable Sensors for Digital Health Technology: A Review. *Sensors 2023, Vol. 23, Page 9498, 23*(23), 9498. DOI: 10.3390/s23239498

Shapiro, B., Barlow, D., & Keller, M. (2022). Autonomous vehicles: The new frontier for in-vehicle advertising and media content. *Journal of Marketing Technology, 33*(1), 77–92.

Shekerbekova, S., Kamalova, G., Iskakova, M., Aldabergenova, A., Abdykerimova, E., & Shetiyeva, K. (2025). Exploring the use of Artificial Intelligence and Augmented Reality tools to improve interactivity in Physical Education teaching and training methods. *Retos, 66*, 935–949. DOI: 10.47197/retos.v66.113540

Shladover, S. E., Gonder, J., & Dixon, L. (2021). The future of connected and autonomous vehicles: Implications for vehicle safety and transportation efficiency. *Transportation Research Part C, Emerging Technologies, 121*, 92–105.

Simons, H., Jones, S., & Wilkins, D. (2021). Consumer adoption of autonomous vehicles: Overcoming trust barriers in marketing strategies. *Journal of Marketing, 85*(6), 23–39.

Singapore, P. U. B. (2020). *Smart Water Management*. https://www.pub.gov.sg

Singer, N. (2019). San Francisco bans facial recognition technology. *The New York Times*. https://www.nytimes.com

Sisi, Z., & Souri, A. (2021). Blockchain technology for energy-aware mobile crowd sensing approaches in Internet of Things. *Transactions on Emerging Telecommunications Technologies, 35*(4), e4217. Advance online publication. DOI: 10.1002/ett.4217

Skarmeas, D., Saridakis, C., & Leonidou, C. N. (2018). Examining relationship value in cross-border business relationships: A comparison between correlational and configurational approaches. *Journal of Business Research, 89*, 280–286. DOI: 10.1016/j.jbusres.2017.12.039

Skinner, B. F. (2016). *The Technology of Teaching*. BF Skinner Foundation.

Slimi, Z., & Carballido, B. V. (2023). Systematic Review: AI's Impact on Higher Education - Learning, Teaching, and Career Opportunities. *Tem Journal-Technology Education Management Informatics, 12*(3), 1627–1637. https://doi.org/DOI: 10.18421/TEM123-44

Smart Cities Council India. (2020). *Integrated Command and Control Centres: Case Studies.* https://smartcitiescouncil-india.com

Smart Cities Council India. (2021). *Smart City Initiatives and COVID-19 Response.* https://smartcitiescouncil-india.com

Smart Nation Singapore. (2020). *Smart Mobility Initiatives.* https://www.smartnation.gov.sg

Smart Nation Singapore. (2020). Smart Nation Initiatives. Retrieved from https://www.smartnation.gov.sg

Smart Nation Singapore. (2020). *Virtual Singapore Initiative.* https://www.smartnation.gov.sg

Somogyvári, M. (2021) Financial Exclusion in the Digital Payment Space, *Financial and Economic Review*, (65-86)

Soong, R., Pautler, B. G., Moser, A., Jenne, A., Lysak, D. H., Adamo, A., & Simpson, A. J. (2020). CASE (Computer-Assisted Structure Elucidation) Study for an Undergraduate Organic Chemistry Class. *Journal of Chemical Education, 97*(3), 855–860. DOI: 10.1021/acs.jchemed.9b00498

Steenburgh, T. J., Ainslie, A., & Chintagunta, P. K. (2020). Sustainability in marketing: The rise of green consumerism and its effect on the automotive sector. *Journal of Marketing, 84*(6), 22–39.

Sterling, S. (2011). Transformative learning and sustainability: Sketching the conceptual ground. *Learning and teaching in higher education, 5*(11), 17-33.

Su, J., & Yang, W. (2023). Unlocking the Power of ChatGPT: A Framework for Applying Generative AI in Education. *ECNU Review of Education, 6*(3), 355–366. DOI: 10.1177/20965311231168423

Sultan, D., Omarov, B., Rakhymzhanov, A., Niyazov, A., & Baikuvekov, M., B. O. (2024). Development of an artificial intelligence-enabled non-invasive digital stethoscope for monitoring the heart condition of athletes in real-time. *Retos: Nuevas Tendencias En Educación Física. Deporte y Recreación, 60*, 1169–1180.

Sustainable Travel International. (2023). *The sustainable traveler report: Insights from the global eco-conscious tourist.* https://sustainabletravel.org/resources

Tavares, S., & Ferrara, E. (2023). Fairness and Bias in Artificial Intelligence: A Brief Survey of Sources, Impacts, and Mitigation Strategies. *Sci 2024, Vol. 6, Page 3, 6*(1), 3. DOI: 10.3390/sci6010003

Teicholz, P. (2013). BIM for facility managers. *Journal of Building Performance*, *4*(1), 17–28.

Tello, A. (2023). On Digital Colonialism: Data, Algorithms, and the Technological Coloniality of Power in the Global South in *Inmediaciones de la Comunicación*, Vol. 18, no. 2.https://doi.org/DOI: 10.18861/ic.2023.18.2.3523

The Hindu. (2019). Delhi police use facial recognition to identify missing children. https://www.thehindu.com

Tiernan, P., Costello, E., Donlon, E., Parysz, M., & Scriney, M. (2023). Information and Media Literacy in the Age of AI: Options for the Future. *Education Sciences*, *13*(9), 906. DOI: 10.3390/educsci13090906

Tilbury, D. (2011). Education for sustainable development: An expert review of processes and learning.

Tilbury, D. (2011). Higher education for sustainability: a global overview of commitment and progress. *Higher education in the world, 4*(1), 18-28.

Torelli, R., Balluchi, F., & Lazzini, A. (2020). Greenwashing and environmental communication: Effects on stakeholders' perceptions. *Business Strategy and the Environment*, *29*(2), 407–421. DOI: 10.1002/bse.2373

Townsend, A. M. (2013). *Smart Cities: Big Data, Civic Hackers, and the Quest for a New Utopia*. W. W. Norton & Company.

Transport & Environment. (2021). *Flying into trouble: Ryanair's misleading emissions claims*. https://www.transportenvironment.org/publications

Tundrea, E. (2020). Artificial Intelligence in Higher Education: Challenges and Opportunities. *14th International Technology, Education and Development Conference (Inted2020),* 2041–2049. https://doi.org/DOI: 10.21125/inted.2020.0644

Ubago Jiménez, J. L., Puertas Molero, P., González Valero, G., Melguizo Ibáñez, E., Valverde Janer, M., & Ortega Caballero, M. (2022). Uso de los dispositivos móviles (Mhealth) en la práctica deportiva en adolescentes. *Revista INFAD de Psicología. International Journal of Developmental and Educational Psychology.*, *1*(1), 165–170. DOI: 10.17060/ijodaep.2022.n1.v1.2340

UK Advertising Standards Authority (ASA). (2020). *Ryanair low emissions advertising ruling*. https://www.asa.org.uk/rulings/ryanair.html

UK Government. (2020). *Road to zero: Next steps towards cleaner road transport and delivering our industrial strategy*. UK Government.

UNESCO. (2020). *Education for Sustainable Development: A roadmap.* United Nations Educational, Scientific and Cultural Organization. https://unesdoc.unesco.org/ark:/48223/pf0000374802

UNESCO. (2022). *UNESCO declares environmental education must be a core curriculum component by 2025.* Unesco.org. https://www.unesco.org/en/articles/unesco-declares-environmental-education-must-be-core-curriculum-component-2025

UN-Habitat. (2020). *Resilient Cities and Climate Action.* https://unhabitat.org

UNICEF. (2021). *Reimagining Education in Latin America and the Caribbean.* https://www.unicef.org/lac/en/reimagine-education-latin-america-and-caribbean

United Nations. (2018). World Urbanization Prospects: The 2018 Revision. Retrieved from https://population.un.org/wup/

University of Bielefeld/Oberstufenkolleg (1999). The German experience: from ecological education to education for sustainable development. Environment and Development 15 (n4) (pp 60-65) (ISSN 0716 - 1476)

Ushakov, A. (2017). Learning Content Management Systems in Flt: Canadian Experience. *International Journal of English Language and Literature Studies, 6*(1), 25–32. RePEc: asi: ijells:v:6: y:2017: i:1:p:25-32:id:531

Vaca-Trigo, I., & Valenzuela, M. E. (2022). Digitalization of Women in Latin America and the Caribbean: Urgent Action for a Transformative and Equal Recovery (No. 47940). United Nations Economic Commission for Latin America and the Caribbean (ECLAC). RePEc: ecr: col022:47940

Valadares, G. C., de Rezende, D. C., Carvalho, J. F. S., dos Reis, N. D., Evangelista, J. M. F., & Costa, A. P. (2023). Smartphone use and purchasing behavior in supermarkets: An approach based on Practice Theory. *Revista da UI_IPSantarém, 11*(4), 64–76.

Van Winden, W., van den Buuse, D., & van Dijck, E. (2016). Smart city pilot projects: Exploring the challenges and lessons learned. *Journal of Urban Technology, 23*(4), 3–20.

Van Zoonen, L. (2016). Privacy concerns in smart cities. *Government Information Quarterly, 33*(3), 472–480. DOI: 10.1016/j.giq.2016.06.004

Varela-Losada, M., Lorenzo-Rial, M. A., Pérez-Rodríguez, U., & Vega-Marcote, P. (2024). Education for Sustainability in the minimum teaching of Primary Education. *Didáctica de las ciencias experimentales y sociales*, (46), 89-106.

Vega-Marcote, P., Varela-Losada, M., & Álvarez-Suárez, P. (2015). Evaluation of an educational model based on the development of sustainable competencies in basic teacher training in Spain. *Sustainability (Basel)*, *7*(3), 2603–2622. DOI: 10.3390/su7032603

Vega-Ramírez, L., Notario, R. O., & Ávalos-Ramos, M. A. (2020). The Relevance of Mobile Applications in the Learning of Physical Education. *Education Sciences*, *10*(11), 329. DOI: 10.3390/educsci10110329

Vermesan, O., & Friess, P. (2021). The future of autonomous vehicles: Privacy and data ethics. *Journal of Ethics in Information Technology*, *19*(2), 140–152.

Vesna, L. (2025). Digital Divide in AI-Powered Education: Challenges and Solutions for Equitable Learning. *Journal of Information Systems Engineering & Management*, *10*(21s), 300–308. DOI: 10.52783/jisem.v10i21s.3327

Villaverde Castro, J., & Maza Fernández, A. J. (2017). Internal Migration in Spain, 2000–2014: Evidence from a Spatial Econometrics Analysis.

Villavicencio-Cedeño, M. M., Bowen-Anchundia, M. M., Jurado-Martínez, M. G., & Roger-Martínez, I. (2024). The Digital Divide in Student Education: Determining Factors, Educational Consequences, and Proposals for Its Mitigation. *MQRInvestigar*, *8*(4), 6641–6673. DOI: 10.56048/MQR20225.8.4.2024.6641-6673

Vinothkumar, T., Sivaraju, S., Thangavelu, A., & Srithar, S. (2023). An energy-efficient and reliable data gathering infrastructure using the internet of things and smart grids. *Automatika (Zagreb)*, *64*(4), 720–732. DOI: 10.1080/00051144.2023.2205724

Vistorte, A. O. R., Deroncele-Acosta, A., Ayala, J. L. M., Barrasa, A., López-Granero, C., & Martí-González, M. (2024). Integrating artificial intelligence to assess emotions in learning environments: A systematic literature review. *Frontiers in Psychology*, *15*, 1387089. DOI: 10.3389/fpsyg.2024.1387089 PMID: 38966729

Voigt, P., & Von dem Bussche, A. (2017). *The EU General Data Protection Regulation (GDPR): A Practical Guide*. Springer. DOI: 10.1007/978-3-319-57959-7

Wadhwa, S., Rani, S., Kavita, N., Verma, S., Shafi, J., & Wozniak, M. (2022). Energy Efficient Consensus Approach of Blockchain for IoT Networks with Edge Computing. *Sensors (Basel)*, *22*(10), 3733. DOI: 10.3390/s22103733 PMID: 35632142

Walker, J. (1998). *The Viable Systems Model a guide for co-operatives and federations*. ICOM (Industrial Common Ownership Movement). https://metaphorum.org/wp-content/uploads/2022/11/vsm-coop-walker.pdf

Walker, K., & Wan, F. (2012). The harm of symbolic actions and green-washing: Corporate actions and communications on environmental performance and their financial implications. *Journal of Business Ethics*, *109*(2), 227–242. DOI: 10.1007/s10551-011-1122-4

Wang, S., Yu, H., Hu, X., & Li, J. (2020). Participant or Spectator? Comprehending the Willingness of Faculty to Use Intelligent Tutoring Systems in the Artificial Intelligence Era. *British Journal of Educational Technology*, *51*(5), 1657–1673. DOI: 10.1111/bjet.12998

Wang, T., & Cheng, E. C. K. (2021). An Investigation of Barriers to Hong Kong K-12 Schools Incorporating Artificial Intelligence in Education. *Computers and Education. Artificial Intelligence*, *2*, 100031. DOI: 10.1016/j.caeai.2021.100031

Wang, Y., & Wang, X. (2024). Artificial intelligence in physical education: Comprehensive review and future teacher training strategies. *Frontiers in Public Health*, *12*, 1484848. Advance online publication. DOI: 10.3389/fpubh.2024.1484848 PMID: 39583072

Warschauer, M. (2004). *Technology and social inclusion: Rethinking the digital divide*. MIT press.

Watters, A. (2017). Dunce's App: How Silicon Valley's brand of behaviorism has entered the classroom. *The Baffler,* Weston-Sementelli, J., Allen, L. K., & McNamara, D. S. (2018). Comprehension and Writing Strategy Training Improves Performance on Content-Specific Source-Based Writing Tasks. *International Journal of Artificial Intelligence in Education*, *28*(1), 106–137. DOI: 10.1007/s40593-016-0127-7

Weis, T., Zhao, Y., Tsourdos, A., & Yu, J.-E. (2022). Exploration of Educational Possibilities by Four Metaverse Types in Physical Education. *Technologies 2022, Vol. 10, Page 104, 10*(5), 104. DOI: 10.3390/technologies10050104

Wiener, N. (2019). *Cybernetics Or Control and Communication in the Animal and the Machine*. MIT Press., DOI: 10.7551/mitpress/11810.001.0001

Woodruff, K., Hutson, J., & Arnone, K. (2023). *Perceptions and Barriers to Adopting Artificial Intelligence in K-12 Education: A Survey of Educators in Fifty States*. Reimagining Education - The Role of E-Learning, Creativity, and Technology in the Post-Pandemic Era., DOI: 10.5772/intechopen.1002741

World Bank. (2016). *Reducing Non-Revenue Water*. https://www.worldbank.org

World Bank. (2016). *World Development Report 2016: Digital Dividends*. https://www.worldbank.org/en/publication/wdr2016

World Bank. (2021). Digital economy for Africa initiative. https://www.worldbank.org/en/programs/all-africa-digital-transformation

World Bank. (2021). *Global Findex Database 2021: Financial Inclusion, Digital Payments, and Resilience in the Age of COVID-19*. Retrieved from https://www.worldbank.org

World Health Organization. (2019). *Para crecer sanos, los niños tienen que pasar menos tiempo sentados y jugar más.*

Xiang, Z., Zheng, Y., He, M., Shi, L., Wang, D., Deng, S., & Zheng, Z. (2021). Energy-effective artificial internet-of-things application deployment in edge-cloud systems. *Peer-to-Peer Networking and Applications*, *15*(2), 1029–1044. DOI: 10.1007/s12083-021-01273-5

Yang, Y., Chen, L., He, W., Sun, D., & Salas-Pilco, S. (2024). Artificial Intelligence for Enhancing Special Education for K-12: A Decade of Trends, Themes, and Global Insights (2013–2023). *International Journal of Artificial Intelligence in Education.* Advance online publication. DOI: 10.1007/s40593-024-00422-0

Yin, R. K. (2018). *Case study research and applications: Design and methods* (6th ed.). SAGE Publications.

Yunga, F., Tapia Morquecho, C. A., Guerrero Riofrío, P. Y., & Flores Chamba, J. E. (2023). The Effect of Technology on Income Inequality: Implications of the Digital Divide — Evidence for OECD Member Countries. *Accounting and Management*, *68*(1), 260–288. DOI: 10.22201/fca.24488410e.2023.3308

Zanella, A., Bui, N., Castellani, A., Vangelista, L., & Zorzi, M. (2014). Internet of Things for Smart Cities. *IEEE Internet of Things Journal*, *1*(1), 22–32. DOI: 10.1109/JIOT.2014.2306328

Zawacki-Richter, O., Marín, V. I., Bond, M., & Gouverneur, F. (2019). Systematic Review of Research on Artificial Intelligence Applications in Higher Education – Where Are the Educators? *International Journal of Educational Technology in Higher Education*, *16*(1), 1–27. DOI: 10.1186/s41239-019-0171-0

Zha, H., Li, W., Wang, W., & Xiao, J. (2025a). The Paradox of AI Empowerment in Primary School Physical Education: Why Technology May Hinder, Not Help, Teaching Efficiency. *Behavioral Sciences (Basel, Switzerland)*, *15*(2), 240. DOI: 10.3390/bs15020240 PMID: 40001871

Zhao, Z., & Xing, X. (2024). Design of an AI-Enhanced Medical Research-Informed Multi-Sensor-Based Online Physical Education Motion Capture System. *Computer-Aided Design and Applications*, 253–267. DOI: 10.14733/cadaps.2024.S24.253-267

Zhou, T., Wu, X., Wang, Y., Wang, Y., & Zhang, S. (2024). Application of artificial intelligence in physical education: A systematic review. *Education and Information Technologies, 29*(7), 8203–8220. DOI: 10.1007/s10639-023-12128-2

Zhu, S., Ota, K., & Dong, M. (2022). Energy-Efficient artificial intelligence of things with intelligent edge. *IEEE Internet of Things Journal, 9*(10), 7525–7532. DOI: 10.1109/JIOT.2022.3143722

Zsarnoczky-Dulhazi, F., Agod, S., Szarka, S., Tuza, K., & Kopper, B. (2024). AI BASED MOTION ANALYSIS SOFTWARE FOR SPORT AND PHYSICAL THERAPY ASSESSMENT. *Revista Brasileira de Medicina Do Esporte, 30*. https://doi.org/DOI: 10.1590/1517-8692202430012022_0020I

Zugaza, L. A., & Ramos, J. M. L. (2007). Corruption, Development Aid, Poverty, and Human Development. *Economic Bulletin of ICE*, (2917).

# About the Contributors

**Alberto Tomás Delso Vicente** is a Professor in the Department of Business at Rey Juan Carlos University (Spain). Ph.D. in Business with a focus on mobility and the management of electric vehicles, emphasizing legitimacy, uncertainty, and innovation in consumer behavior. Holds an MBA and a degree in Chemical Engineering. Research interests include Marketing, Economic Policy, Behavioral Finance, Sustainable Mobility, Renewable Energies, and Innovation Management. Expert in the use of Structural Equation Modeling (SEM) methodologies with tools like PLS and AMOS. Member of various academic and industry-related research groups and leader in international academic projects and Erasmus+ initiatives. Active participant in conferences and collaborations that bridge academic knowledge with practical industry applications. Additionally, he is a member of the European Academy of Management and Business Economics (AEDEM).

**María García de Blanes Sebastián** is a distinguished academic with a Ph.D. in Business Economics and a strong track record in research focused on technological innovation, sustainability, and technology acceptance. Below is a selection of her most relevant publications, which support her suitability as a co-editor for the book "Innovation and Sustainability in Electric and Autonomous Mobility"

**Oscar Aguado-Tevar**, Executive Manager at Universidad Nebrija, a Mining Engineer, and holds a Ph.D. from UPM. He also has a Bachelor's degree in Business Administration from HEC-Paris and earned two Master's degrees in Energy Economics from the French Institute of Petroleum and the University of Oklahoma. In addition to his management role at Nebrija, he combines teaching responsibilities at Nebrija with leading the Economics module of the Master's in Oil & Gas (ETSIME-UPM). He has taught finance courses at CUNEF and supervised numerous Master's theses. His management experience includes various positions at Nebrija, CUNEF, and UPM, as well as participation in accreditation panels such as ABET and EFMD/ENQHEEI. Furthermore, he has extensive professional experience in the banking sector, having worked for 16 years at BBVA: 9 years in Strategy and Business Development for Asset Management and 7 years in wholesale banking within the energy sector.

**Luis Díaz-Marcos**, "CEO of Nebrija University, where he has previously been Vice Chancellor of Academic Management and Teaching Staff. Professional with more than 30 years of academic-executive experience. Graduate in Law (UCM Madrid), Master in Business Administration (University of Houston, USA), PhD in Business Administration (Nebrija University, Madrid), has developed most of his professional career in top-level executive positions in academic positions in various institutions: leading new projects, growing the business and strengthening the development of academic and institutional activity in the national and international arena. Along with his current responsibilities at Nebrija University as General Manager, he continues to develop his teaching and research activities in the areas of strategic management, corporate governance and sustainability policies."

**Margarita Núñez-Canal**, Professor and researcher in Business Administration at Nebrija Business and Technology School, Universidad Nebrija. Accredited as "Profesor titular" with one recognized research period (sexenio). Senior Fellow of the Higher Education Academy. Holds a PhD. At Universidad San Pablo CEU, Master's in Law and Economics from the University of Rotterdam and a Law degree from the Complutense University of Madrid. Researcher specializing in business, entrepreneurship, digital skills, artificial intelligence in education, and digitalization. Currently serves as academic Director at Nebrija Business and Technology School.

**Alberto Azuara**, PhD in Tourism from Universidad Rey Juan Carlos (Spain) since 2018. Bachelor's Degree in Tourism from Universidad Autónoma de Madrid, and Master's Degree in International Tourism Management from Universidad Rey Juan Carlos. He conducts multidisciplinary research in various areas related to tourism. His two main lines of research focus on the management of cultural destinations in rural areas, as well as the analysis of tourist behavior and the adoption and use of new tourism technologies. He has published in indexed journals listed in JCR-WOS and SJR-Scopus, as well as in high-quality journals included in ESCI, Latindex, among others. He has also authored several book chapters published by prestigious national and international publishers. He has participated as a speaker at renowned conferences and congresses and has collaborated as a researcher in a regional research project. Regarding teaching experience, he has been teaching since the 2018/2019 academic year in various undergraduate, double degree, master's, and doctoral programs, serving as the lead instructor for several courses. He has accumulated over 1,000 hours of undergraduate teaching and more than 70 hours at the master's level. He was awarded one teaching merit in the 2021 DOCENTIA program, with an “Excellent” rating. In terms of academic management, he currently serves as Academic Exchange Coordinator for the Erasmus+ program, Deputy Coordinator of the Bachelor's Degree in Tourism, and Coordinator of the Blended Learning Bachelor's Degree in Tourism at URJC.

**Amutha B** received the B.E. in Electronics and Instrumentation Engineering from Annamalai University, Chidambaram and M.E. in Computer Science and Engineering from SRM Institute of Science and technology Chennai, India, Currently she is an professor at Department of Computing Technologists Institute of Science and technology, Chennai India .Her Current Research interests include 5G,Cognitive Radio,SDN, Internet of Things,Wireless Computing,smart cities

**Antonio Pérez de Juan** is a researcher specializing in digital transformation and urban mobility. He focuses on applying machine learning to analyze socioeconomic patterns in payment behaviors.

**Elena Fernández** is professor at Facultad de Ciencias de la Economía y la Empresa, Universidad Rey Juan Carlos. She has worked for 25 years in the Spanish publishing industry, in different positions. She is PhD student, and her research interests lie in teams and behavioral economics.

**Felipe García-Soriano** is a sociologist and PhD candidate in the Economy, Business, and Society program at the University of Alicante. His academic work focuses on systems thinking, organizational cybernetics, and complexity theory, with a particular interest in their application to participatory governance and social innovation. He holds a Master’s degree in Social Innovation and Change Dynamics (UA) and a Bachelor’s degree in Sociology from the University of Chile. His research explores how cyber-systemic frameworks—especially the Viable System Model—can be used to analyze and strengthen decentralized organizations. He has lectured at Tampere University (Finland) in the Master’s in Game Studies program, teaching the design of role-playing games as tools for systemic learning and organizational transformation. Felipe also collaborates with the University of Alicante’s Research Group on Innovation and Social Change, bridging theoretical insight with practical methodologies for managing complexity in social systems.

**Raúl Gómez Martínez** is a professor at Universidad Rey Juan Carlos specializing in data science and artificial intelligence. His research focuses on machine learning applications in social and economic contexts.

**J. Sudarvel** working as an Assistant Professor of Management at Karpagam Academy of Higher Education. He produced five Ph.D. Scholars and four Ph.D. Scholars are under his guidance. Further, to his credit he published more than five articles in Scopus indexed journals and published more than 50 articles in international journals. He is acting as a Subject Experts in Board of studies of various autonomous colleges.

**K. Jothi** working as an Associate Professor and Head, Department of Commerce with E Commerce at KPR College of Arts, Science and Research, Coimbatore. She produced more than 10 Ph.D. Scholars and published more than 10 articles in Scopus Indexed Journals and more than 50 articles in International journals. She is also acting as a subject expert in Board of Studies at various colleges.

**Rubén Madrigal-Cerezo**, Academic Coordinator of the Master's Degree in Teacher Training at UNIE Universidad. His research focuses on the integration of artificial intelligence and digital competencies in education. He has co-authored studies on technology-enhanced pedagogy in physical education, emphasizing student engagement and learning outcomes. His work addresses challenges such as digital inequality and the need for teacher training in technological tools. Madrigal-Cerezo advocates for responsible and inclusive adoption of AI in educational settings

**Alexandra Martín-Rodríguez**, PhD in Sport Sciences, is a lecturer and researcher at Universidad Europea de Madrid and UNIE. Her work explores the interplay between physical activity, nutrition, and psychological well-being, especially in youth development. She has published extensively on physical activity, ADHD, and tech-enhanced education in health contexts.

**María Luisa Medrano García** is an expert in sustainability and environmental management. She focuses on integrating sustainable practices within urban development and policy frameworks.

**Rubén Nicolás Sans** serves as the Dean of the School of Engineering, Science & Technology at UNIE and hold the title of Associate Professor (Prof. Titular de Universidad, ANECA 2023), with an accredited six-year research period (Sexenio CNEAI-ANECA 2021). He earned his PhD in Artificial Intelligence in 2014 and was recognized as a Senior Fellow of the Higher Education Academy in 2018. He has published 34 articles in indexed research journals band authored 18 book chapters. In 2009, He was a visiting scholar at Carnegie Mellon University's Human Sensing Lab in Pittsburgh, PA, USA. In 2022, He completed a postdoctoral research visit at Universidad del Atlántico Medio. Currently, He serve as Principal Investigator of the Research Group in Generative AI in Education.

**R. Velmurugan** presently working as an Associate Professor in Commerce, Karpagam Academy of Higher Education, Coimbatore. He started his career as a Project Fellow under UGC Sponsored Major Research Project at NGM College, Pollachi. He has 18 years of teaching experience has 15 years of research experience. To his credit he produced 22 Ph.D. Scholars and 5 M.Phil. Scholars. At present 1 scholar submitter his thesis, 6 Ph.D. Scholars are under his guidance. Moreover, to his credit he published 168 articles in various International Journals, of which 82 articles are published in Scopus Indexed Journals and 15 articles in Web of Science and 76 articles in edited volumes. He also participates and presented 27 papers in various International Seminar / Conference and 54 National Seminar / Conference. He acted as an external examiner for various M.Phil. and Ph.D. Viva voce examinations. He acted as a Vice Chancellor nominee for Doctoral Committee Expert at Bharathiar University, Coimbatore and Periyar University, Salem. He acted as a Chairperson at various National Level Conferences. He also acted as an Expert for Syllabus Scrutiny Committee, Question Paper Scrutiny Committee and Expert members for Board of Studies at various reputed Colleges. Considering his meritorious contribution, he received awards like Best Paper Presenter, Best Research Supervisor and Best Faculty Awards from various research foundations and councils. He too acted as a resource person for 27 Guest Lectures, 30 Workshops and 10 faculty development programmes.

**Delsi Robinsha S** received the B.E. in computer science and Engineering in SRM Valliammai Engineering College and M.Tech in Computing Technologies from SRM Institute of Science and technology Chennai, India in 2020 and 2022 Currently Pursuing Ph.D. in Computing Technologies in SRM Institute of Science and technology, Chennai India. Her current research interests are Internet of Things and Machine Learning .

**José Ramón Sarmiento Guede**, PhD in Marketing from Universidad Rey Juan Carlos, Master's Degree in Marketing Management from ESIC Business & Marketing School, Master's Degree in Sales Management from Universidad Rey Juan Carlos, and Master's Degree in Tourism Business Management from ESERP Business School. Accredited by ANECA since June 2018 as Assistant Professor, Private University Professor, and Tenured Lecturer. He specializes in Relationship, Direct and Interactive Marketing, Digital Marketing Strategies, Digital Communication, and Social Media Marketing.

**Ravi Thirumalaisamy** working as an Assistant Professor, Department of Business and Economics, Modern College of Business and Science. To his credit, he published more than 10 articles in Scopus indexed journals.

**Enriqueta Villanueva Montero** holds a degree in Economics from the University of Castilla-La Mancha and a PhD cum laude in Economics and Business from the same university. Professionally, she has worked as a freelance consultant, head of the Tax and Accounting Department. She is also a freelance economist, insolvency administrator, and consultant in the field of business creation, organization, and management. At the teaching level, she has taught seminars and specialized courses related to her career path, specializing in Taxation, Finance, and Audit Management. She has taught master's and postgraduate courses at private universities and business schools, participating in their academic management. She has also been a tutor and member of the Master's Thesis Panel.

# Index

## A



## B



## C



## D



## E

## F



## G



## I



## M



## N



## P



## R



## S

## T



## U



## V

www.ingramcontent.com/pod-product-compliance
Lightning Source LLC
LaVergne TN
LVHW080857300926
849162LV00003B/445
* 9 7 9 8 3 3 7 3 2 8 0 2 7 *